The Analysis of Sensations, and the Relation of the Physical to the Psychical

By Dr. Ernst Mach

Emeritus Professor at the University of Vienna

Translated by C. M. Williams

Revised and Supplemented from the fifth German Edition by Sydney Waterlow

PANTIANOS
CLASSICS

Published by Pantianos Classics

ISBN-13: 978-1-78987-488-4

First published in 1897

This translation to English was first published in 1914

Contents

Translator's Note

This book is not so much a new edition of the English translation of Professor Mach's *Contributions to the Analysis of the Sensations,* which was published in 1897, [1] but is, as the more comprehensive title indicates, an almost entirely new book. The *Contributions* originally appeared in 1886. The English translation of 1897 contained a certain amount of new matter, most of which was embodied by Professor Mach in his second edition (1900). Since then there have been three more editions: two of them, the third and the fifth, containing important changes and additions of such extent that the fifth edition, of which this translation is now offered to the English-speaking public, is a book nearly twice as long as the original English translation.

It may therefore be convenient to mention here the principal respects in which this book differs from the translation of 1897. Six chapters are entirely new, namely, Chapter III., on "My Relation to Richard Avenarius and other Thinkers"; Chapter V., on "Physics and Biology: Causality and Teleology"; Chapter VIII., on "The Will"; Chapter IX., on "Biologico-teleological Considerations as to Space"; Chapter XI., on "Sensation, Memory and Association," and Chapter XV., on "How my Views have been received." Further, the eight chapters of the original edition have all been greatly expanded. Chapter II. now contains most of the matter which appeared as an appendix to the translation of 1897. Chapter VII. contains six sections by Dr. Josef Pollak on recent research as to the functions of the labyrinth of the ear.

It will be seen that the changes and additions fall, on the whole, into two classes. They are made with the object either of amplifying and bringing up to date the author's original discussions of points of detail, or of explaining and justifying his more general views as to the relation between different branches of science and as to questions on the borderland between science and philosophy. It ill becomes a translator to indulge the temptation, which he yet must feel, to turn commentator or eulogist; but I may perhaps be allowed to point out the great interest attaching to the explanations here given by the veteran physicist and philosopher (if Professor Mach will allow the word "philosopher") of the way in which his views were developed.

For those parts of the text which are identical with the English edition of 1897 I have availed myself largely of Miss Williams's excellent translation. Finally, I must add that the whole of the present translation has been most kindly read, in manuscript, by Professor Mach himself.

SYDNEY WATERLOW

[1] Contributions to the Analysis of the Sensations, by Dr. Ernst Mach. Translated by C. M. Williams. Chicago: Open Court Publishing Co., 1897.

Author's Prefaces

Preface to the Fifth Edition

THE text of this edition has been enlarged by a number of new passages and notes. There is an insertion of some length on recent investigations as to the sense of orientation; this is from the pen of Professor Josef Pollak, who has also been so kind as to read the proofs of the whole book and to correct the index. For all these services I owe him my heartiest thanks. A mistake as to Ewald's theory of audition has been corrected. I have noted with satisfaction that a view of the relation between the physical and the psychical, which is almost identical with the view advocated here, occurs in a book by Alfred Binet (*L'Âme et le Corps,* Paris, 1905).

VIENNA, *May* 1906

Preface to the First Edition

THE frequent excursions which I have made into this province have all sprung from the profound conviction that the foundations of science as a whole, and of physics in particular, await their next greatest elucidations from the side of biology, and especially from the analysis of the sensations.

I am aware, of course, that I have succeeded in contributing but little to the attainment of this end. The very fact that my investigations have been carried on, not in the way of a profession, but only at odd moments, and frequently only after long interruptions, must detract considerably from the value of my scattered publications, or perhaps even lay me open to the silent reproach of desultoriness. So much the more, therefore, am I under especial obligations to those investigators, such as E. Hering, V. Hensen, W. Preyer and others, who have directed attention either to the matter of my writings or to my methodological expositions.

The present compendious and supplementary presentation of my views will, perhaps, place my attitude in a somewhat more favorable light, for it will be seen that in all cases I have had in mind the same problem, no matter how varied or numerous were the single facts investigated. Although I can lay no claim whatever to the title of physiologist, and still less to that of philosopher, yet I venture to hope that the work thus undertaken, purely from a strong desire for self-enlightenment, by a physicist unconstrained by the conventional barriers of the specialist, may not be entirely without value for others also, even though I may not be everywhere in the right.

My natural bent for the study of these questions received its strongest stimulus twenty-five years ago from Fechner's *Elemente der Psychophysik* (Leipzig, 1860), but my greatest assistance was derived from Bering's solution of the two problems referred to on pages 69 and 168.

To readers who, for any reason, desire to avoid more general discussions, I recommend the omission of the first and last chapters. For me, however, the conception of the whole and the conception of the parts are so intimately related that I should scarcely be able to separate them.

PRAGUE, *November* 1885

Preface to the Second Edition

THIS book was intended to have the effect of an *aperçu,* and, if I may judge from the occasional utterances of Avenarius, H. Cornelius, James, Külpe, Loeb, Pearson, Petzoldt, Willy, and others, it seems to have fulfilled its object. It now appears, after fourteen years, in a new edition. This is a rather bold undertaking. For to allow the book to swell out into a bulky volume, by adding accounts of many experimental researches on points of detail, and by noticing at length the literature which has appeared since it was first published, would not be in keeping with its character. Yet I was unwilling to let slip this last opportunity without once again saying something on a subject which I have so much at heart. I have therefore added the supplements and elucidations most urgently required, principally by inserting short chapters in the original text. One of these, the second, has already appeared in the English edition published in 1897.

One and the same view underlies both my epistemologico-physical writings and my present attempt to deal with the physiology of the senses the view, namely, that all metaphysical elements are to be eliminated as superfluous and as destructive of the economy of science. If I have not entered in these pages upon a detailed critical and polemical discussion of views that are opposed to my own, this is in truth not from contempt of my opponents, but because I am convinced that questions of this kind cannot be decided by controversies and dialectic combats. The only really profitable course is to carry one's half-thought or, it may be, one's paradoxical idea patiently about with one for years, and to make an honest effort to complete the half-thought, or to strip away the paradoxical element, as the case may be. Those readers who, after turning over the first pages, lay the book aside, because their convictions are such that they cannot follow me any further, will only be doing exactly what I myself have sometimes been compelled to do.

In its former shape the book met with much friendly acceptance, but it also aroused strenuous opposition. Readers who wish to go more deeply into the subjects of which it treats, may find it useful to know that Willy, in a recently published work (*Die Krisis in der Psychologie,* Leipzig, 1899), in which a position closely allied to my own is adopted, opposes my views in many points of detail.

VIENNA, *April* 1900

Preface to the Third Edition

CONTRARY to my expectation, the second edition was exhausted in a few months. I have not hesitated to make certain additions which may help to put my views in a clearer light, though without altering the text of 1886 in any essential respect. Two passages only of the second edition (paragraph 7, p. 11, and paragraph 11, p. 15) have been cast in a clearer form. Dr. A. Lampa, lecturer in physics in this University, was told by several readers that these passages were often understood in a one-sided idealistic sense, an interpretation which I in no wise intended. I am greatly obliged to Dr. Lampa for giving me this information. Chapters IX. and XV., in which subjects touched upon in the second edition are developed at greater length, are new additions.

Unless all indications are deceptive, I no longer occupy, as regards my views, anything like so isolated a position as I did even a few years ago. In addition to the school of Avenarius, there are also younger thinkers, such as H. Gomperz, who are approaching my point of view by their own paths. The differences that still remain over seem to me not irreconcilable. But it would be premature to dispute about them yet. "But the question is one in which it is peculiarly difficult to make out precisely what another man means, and even what one means oneself." The author of this delightfully humorous remark was W. K. Clifford, the mathematician ("On the Nature of Things-in-themselves," *Lectures,* vol. ii. p. 88), a writer with an extremely close affinity to myself in the direction of his thought.

VIENNA, November 1901

Preface to the Fourth Edition

THE opinion, which is gradually coming to the front, that science ought to be confined to the compendious representation of the actual, necessarily involves as a consequence the elimination of all superfluous assumptions which cannot be controlled by experience, and, above all, of all assumptions that are metaphysical in Kant's sense. If this point of view is kept firmly in mind in that wide field of investigation which includes the physical and the psychical, we obtain, as our first and most obvious step, the conception of the sensations as the common elements of all possible physical and psychical experiences, which merely consist in the different kinds of ways in which these elements are combined, or in their dependence on one another. A whole series of troublesome pseudo-problems at once disappears. The aim of this book is not to put forward any system of philosophy, or any comprehensive theory of the universe. It is only the consequences of this single step, to which any number of others may be attached, that are examined here. An attempt is made, not to solve all problems, but to reach an epistemological position which shall prepare the way for the co-operation of special departments of research, that are widely removed from one another, in the solution of important problems of detail.

It is from this point of view that the accounts of special investigations, which are given here, should be regarded. If there is no essential difference between the physical and the psychical, we shall hope to trace the same exact connexion, which we seek in everything that is physical, in the relations between the physical and the psychical also. We then expect to find, corresponding to all the details which physiological analysis can discover in the sensations, as many details of physical nerve-process. I have tried to describe this relation, so far as I have been able to do so,

Expressions of extravagant praise and of equally extravagant blame, have come to my ears. I hope that what I have just said, by moderating both, may promote a sober judgment. When, about thirty-five years ago, I succeeded, by overcoming my own prejudices, in firmly establishing my present position and in setting myself free from the greatest intellectual discomfort of my life, I attained thereby to a certain satisfaction. At that time I was only acquainted with Kant and Herbart. To-day I see that a whole host of philosophers positivists, critical empiricists, adherents of the philosophy of immanence, and certain isolated scientists as well have all, without any knowledge of one another's work, entered upon paths which, in spite of all their individual differences, converge almost towards one point. If, in these circumstances, I cannot rate very high the value of my individual labours, I may nevertheless be permitted to believe that I have not merely pursued a subjective phantom, but have contributed towards the attainment of a goal at which many others besides myself have been aiming. It would of course be absurd, where ideas are concerned of which the leading threads reach back to antiquity, to set up any claims to priority.

Dr. Josef Pollak and Dr. Wolfgang Pauli, lecturers in the Faculty of Medicine in the University of Vienna, have been so extremely kind as to read the proofs, for which I thank them most heartily.

VIENNA, *November* 1902

I. Introductory Remarks: Antimetaphysical

1.

THE great results achieved by physical science in modern times results not restricted to its own sphere but embracing that of other sciences which employ its help have brought it about that physical ways of thinking and physical modes of procedure enjoy on all hands unwonted prominence, and that the greatest expectations are associated with their application. In keeping with this drift of modern inquiry, the physiology of the senses, gradually abandoning the method of investigating sensations in themselves followed by men like Goethe, Schopenhauer, and others, but with greatest success by Johannes Müller, has also assumed an almost exclusively physical character. This tendency must appear to us as not altogether appropriate, when we reflect that physics, despite its considerable development, nevertheless constitutes but a portion of a *larger* collective body of knowledge, and that it is unable, with its limited intellectual implements, created for limited and special purposes, to exhaust all the subject-matter in question. Without renouncing the support of physics, it is possible for the physiology of the senses, not only to pursue its own course of development, but also to afford to physical science itself powerful assistance. The following simple considerations will serve to illustrate this relation between the two.

2.

Colors, sounds, temperatures, pressures, spaces, times, and so forth, are connected with one another in manifold ways; and with them are associated dispositions of mind, feelings, and volitions. Out of this fabric, that which is relatively more fixed and permanent stands prominently forth, engraves itself on the memory, and expresses itself in language. Relatively greater permanency is exhibited, first, by certain complexes of colors, sounds, pressures, and so forth, functionally connected in time and space, which therefore receive special names, and are called bodies. Absolutely permanent such complexes are not.

My table is now brightly, now dimly lighted. Its temperature varies. It may receive an ink stain. One of its legs may be broken. It may be repaired, polished, and replaced part by part. But, for me, it remains the table at which I daily write.

My friend may put on a different coat. His countenance may assume a serious or a cheerful expression. His complexion, under the effects of light or emotion, may change. His shape may be altered by motion, or be definitely changed. Yet the number of the permanent features presented, compared with the number of the gradual alterations, is always so great, that the latter may be overlooked. It is the same friend with whom I take my daily walk.

My coat may receive a stain, a tear. My very manner of expressing this shows that we are concerned here with a sum-total of permanency, to which the new element is added and from which that which is lacking is subsequently taken away.

Our greater intimacy with this sum-total of permanency, and the preponderance of its importance for me as contrasted with the changeable element, impel us to the partly instinctive, partly voluntary and conscious economy of mental presentation and designation, as expressed in ordinary thought and speech. That which is presented in a single image receives a single designation, a single name.

Further, that complex of memories, moods, and feelings, joined to a particular body (the human body), which is called the "I" or "Ego," manifests itself as relatively permanent. I may be engaged upon this or that subject, I may be quiet and cheerful, excited and ill-humored. Yet, pathological cases apart, enough durable features remain to identify the ego. Of course, the ego also is only of relative permanency.

The apparent permanency of the ego consists chiefly in the single fact of its continuity, in the slowness of its changes. The many thoughts and plans of yesterday that are continued to-day, and of which our environment in waking hours incessantly reminds us (whence in dreams the ego can be very indistinct, doubled, or entirely wanting), and the little habits that are unconsciously and involuntarily kept up for long periods of time, constitute the groundwork of the ego. There can hardly be greater differences in the egos of different people, than occur in the course of years in one person. When I recall to-day my early youth, I should take the boy that I then was, with the exception of a few individual features, for a different person, were it not for the existence of the chain of memories. Many an article that I myself penned twenty years ago impresses me now as something quite foreign to myself. The very gradual character of the changes of the body also contributes to the stability of the ego, but in a much less degree than people imagine. Such things are much less analysed and noticed than the intellectual and the moral ego. Personally, people know themselves very poorly. [1] When I wrote these lines in 1886, Ribot's admirable little book, *The Diseases of Personality* (second edition, Paris, 1888, Chicago, 1895), was unknown to me. Ribot ascribes the principal role in preserving the continuity of the ego to the general sensibility. Generally, I am in perfect accord with his views. [2]

The ego is as little absolutely permanent as are bodies. That which we so much dread in death, the annihilation of our permanency, actually occurs in life in abundant measure. That which is most valued by us, remains preserved in countless copies, or, in cases of exceptional excellence, is even preserved of itself. In the best human being, however, there are individual traits, the loss of which neither he himself nor others need regret. Indeed, at times, death, viewed as a liberation from individuality, may even become a pleasant thought. Such reflections of course do not make physiological death any the easier to bear.

After a first survey has been obtained, by the formation of the substance-concepts "body" and "ego" (matter and soul), the will is impelled to a more exact examination of the changes that take place in these relatively permanent existences. The element of change in bodies and the ego, is in fact, exactly what moves the will [3] to this examination. Here the component parts of the complex are first exhibited as its properties. A fruit is sweet; but it can also be bitter. Also, other fruits may be sweet. The red color we are seeking is found in many bodies. The neighborhood of some bodies is pleasant; that of others, unpleasant. Thus, gradually, different complexes are found to be made up of common elements. The visible, the audible, the tangible, are separated from bodies. The visible is analysed into colors and into form. In the manifoldness of the colors, again, though here fewer in number, other component parts are discerned such as the primary colors, and so forth. The complexes are disintegrated into elements, [4] that is to say, into their ultimate component parts, which hitherto we have been unable to subdivide any further. The nature of these elements need not be discussed at present; it is possible that future investigations may throw light on it. We need not here be disturbed by the fact that it is easier for the scientist to study relations of relations of these elements than the direct relations between them.

3.

The useful habit of designating such relatively permanent compounds by single names, and of apprehending them by single thoughts, without going to the trouble each time of an analysis of their component parts, is apt to come into strange conflict with the tendency to isolate the component parts. The vague image which we have of a given permanent complex, being an image which does not perceptibly change when one or another of the component parts is taken away, seems to be something which exists in itself. Inasmuch as it is possible to take away singly every constituent part without destroying the capacity of the image to stand for the totality and to be recognised again, it is imagined that it is possible to subtract all the parts and to have something still remaining. Thus naturally arises the philosophical notion, at first impressive, but subsequently recognized as monstrous, of a "thing-in-itself," different from its "appearance," and unknowable. [5]

Thing, body, matter, are nothing apart from the combilations of the elements, the colors, sounds, and so forth nothing apart from their so-called attributes. That protean pseudo-philosophical problem of the single thing with its many attributes, arises wholly from a misinterpretation of the fact, that summary comprehension and precise analysis, although both are provisionally justifiable and for many purposes profitable, cannot be carried on simultaneously. A body is one and unchangeable only so long as it is unnecessary to consider its details. Thus both the earth and a billiard-ball are spheres, if we are willing to neglect all deviations from the spherical form, and if greater precision is not necessary. But when we are obliged to carry on investigations in orography or microscopy, both bodies cease to be spheres.

4.

Man is pre-eminently endowed with the power of voluntarily and con-sciously determining his own point of view. He can at one time disregard the most salient features of an object, and immediately thereafter give attention to its smallest details; now consider a stationary current, without a thought of its contents (whether heat, electricity or fluidity), and then measure the width of a Fraunhofer line in the spectrum; he can rise at will to the most general abstractions or bury himself in the minutest particulars. Animals possess this capacity in a far less degree. They do not assume a point of view, but are usually forced to it by their sense-impressions. The baby that does not know its father with his hat on, the dog that is perplexed at the new coat of its master, have both succumbed in this conflict of points of view. Who has not been worsted in similar plights? Even the man of philosophy at times succumbs, as the grotesque problem, above referred to, shows.

In this last case, the circumstances appear to furnish a real ground of justi-fication. Colors, sounds, and the odors of bodies are evanescent. But their tangibility, as a sort of constant nucleus, not readily susceptible of annihila-tion, remains behind; appearing as the vehicle of the more fugitive properties attached to it. Habit, thus, keeps our thought firmly attached to this central nucleus, even when we have begun to recognize that seeing hearing, smell-ing, and touching are intimately akin in character. A further consideration is, that owing to the singularly extensive development of mechanical physics a kind of higher reality is ascribed to the spatial and to the temporal than to colors, sounds, and odors; agreeably to which, the temporal and spatial links of colors, sounds, and odors appear to be more real than the colors, sounds and odors themselves. The physiology of the senses, however, demonstrates, that spaces and times may just as appropriately be called sensations as col-ors and sounds. But of this later.

5.

Not only the relation of bodies to the ego, but the ego itself also, gives rise to similar pseudo-problems, the character of which may be briefly indicated as follows:

Let us denote the above-mentioned elements by the letters $A\ B\ C...$, $K\ L\ M...$, $\alpha\ \beta\ \gamma$ ... Let those complexes of colors, sounds, and so forth, commonly called bodies, be denoted, for the sake of clearness, by $A\ B\ C...$; the complex, known as our own body, which is a part of the former complexes distinguished by certain peculiarities, may be called $K\ L\ M...$; the complex composed of voli-tions, memory-images, and the rest, we shall represent by $\alpha\ \beta\ \gamma...$ Usually, now, the complex $\alpha\ \beta\ \gamma...K\ L\ M...$, as making up the ego, is opposed to the complex $A\ B\ C...$, as making up the world of physical objects; sometimes also,

$\alpha \beta \gamma$... is viewed as ego, and $K L M...A B C$...as world of physical objects. Now, at first blush, $A B C$...appears independent of the ego, and opposed to it as a separate existence. But this independence is only relative, and gives way upon closer inspection. Much, it is true, *may* change in the complex $\alpha \beta \gamma$... without much perceptible change being induced in $A B C$...; and *vice versa*. But many changes in $\alpha \beta \gamma$... do pass, by way of changes in $K L M$..., to $A B C$...; and *vice versa*. (As, for example, when powerful ideas burst forth into acts, or when our environment induces noticeable changes in our body.) At the same time the group $K L M$...appears to be more intimately connected with $\alpha \beta \gamma$...and with $A B C$..., than the latter with one another; and their relations find their expression in common thought and speech.

Precisely viewed, however, it appears that the group $A B C$...is always codetermined by $K L M$. A cube when seen close at hand, looks large; when seen at a distance, small; its appearance to the right eye differs from its appearance to the left; sometimes it appears double; with closed eyes it is invisible. The properties of one and the same body, therefore, appear modified by our own body; they appear conditioned by it. But where, now, is that *same* body, which appears so *different?* All that can be said is, that with different $K L M$ different $A B C$... are associated. [6]

A common and popular way of thinking and speaking is to contrast "appearance" with "reality." A pencil held in front of us in the air is seen by us as straight; dip it into the water, and we see it crooked. In the latter case we say that the pencil *appears* crooked, but is in *reality* straight. But what justifies us in declaring one fact rather than another to be the reality, and degrading the other to the level of appearance? In both cases we have to do with facts which present us with different combinations of the elements, combinations which in the two cases are differently conditioned. Precisely because of its environment the pencil dipped in water is optically crooked; but it is tactually and metrically straight. An image in a concave or flat mirror is *only* visible, whereas under other and ordinary circumstances a tangible body as well corresponds to the visible image. A bright surface is brighter beside a dark surface than beside one brighter than itself. To be sure, our expectation is deceived when, not paying sufficient attention to the conditions, and substituting for one another different cases of the combination, we fall into the natural error of expecting what we are accustomed to, although the case may be an unusual one. The facts are not to blame for that. In these cases, to speak of "appearance" may have a practical meaning, but cannot have a scientific meaning. Similarly, the question which is often asked, whether the world is real or whether we merely dream it, is devoid of all scientific meaning. Even the wildest dream is a fact as much as any other. If our dreams were more regular, more connected, more stable, they would also have more practical importance for us. In our waking hours the relations of the elements to one another are immensely amplified in comparison with what they were in our dreams. We recognize the dream for what it is. When the process is reversed, the field of psychic vision is narrowed; the contrast is almost entirely lacking. Where there is no contrast, the distinction between dream and waking, between appearance and reality, is quite otiose and worthless.

The popular notion of an antithesis between appearance and reality has exercised a very powerful influence on scientific and philosophical thought. We see this, for example, in Plato's pregnant and poetical fiction of the Cave, in which, with our backs turned towards the fire, we observe merely the shadows of what passes (*Republic,* vii. 1). But this conception was not thought out to its final consequences, with the result that it has had an unfortunate influence on our ideas about the universe. The universe, of which nevertheless we are a part, became completely separated from us, and was removed an infinite distance away. Similarly, many a young man, hearing for the first time of the refraction of stellar light, has thought that doubt was cast on the whole of astronomy, whereas nothing is required but an easily effected and unimportant correction to put everything right again.

6.

We see an object having a point S. If we touch S, that is, bring it into connexion with our body, we receive a prick. We can see S, without feeling the prick. But as soon as we feel the prick we find S on the skin. The visible point, therefore, is a permanent nucleus, to which the prick is annexed, according to circumstances, as something accidental. From the frequency of analogous occurrences we ultimately accustom ourselves to regard all properties of bodies as "effects" proceeding from permanent nuclei and conveyed to the ego through the medium of the body; which effects we call sensations. By this operation, however, these nuclei are deprived of their entire sensory content, and converted into mere mental symbols. The assertion, then, is correct that the world consists only of our sensations. In which case we have knowledge *only* of sensations, and the assumption of the nuclei referred to, or of a reciprocal action between them, from which sensations proceed, turns out to be quite idle and superfluous. Such a view can only suit with a half-hearted realism or a half-hearted philosophical criticism.

7.

Ordinarily the complex $\alpha\beta\gamma...\ K\ L\ M...$ is contrasted as ego with the complex $A\ B\ C...$ At first only those elements of $A\ B\ C...$that more strongly alter $\alpha\beta\gamma...$, as a prick, a pain, are wont to be thought of as comprised in the ego. Afterwards, however, through observations of the kind just referred to, it appears that the right to annex $A\ B\ C...$to the ego nowhere ceases. In conformity with this view the ego can be so extended as ultimately to embrace the entire world. [7] The ego is not sharply marked off, its limits are very indefinite and arbitrarily displaceable. Only by failing to observe this fact, and by unconsciously narrowing those limits, while at the same time we enlarge them, arise, in the conflict of points of view, the metaphysical difficulties met with in this connexion.

As soon as we have perceived that the supposed unities "body" and "ego" are only makeshifts, designed for provisional orientation and for definite practical ends (so that we may take hold of bodies, protect ourselves against pain, and so forth), we find ourselves obliged, in many more advanced scientific investigations, to abandon them as insufficient and inappropriate. The antithesis between ego and world, between sensation (appearance) and thing, then vanishes, and we have simply to deal with the connexion of the elements $\alpha \beta \gamma...A B C...K L M...$, of which this antithesis was only a partially appropriate and imperfect expression. This connexion is nothing more or less than the combination of the above-mentioned elements with other similar elements (time and space). Science has simply to accept this connexion, and to get its bearings in it, without at once wanting to explain its existence.

On a superficial examination the complex $\alpha \beta \gamma...$appears to be made up of much more evanescent elements than $A B C...$ and $K L M...$, in which last the elements seem to be connected with greater stability and in a more permanent manner (being joined to solid nuclei as it were). Although on closer inspection the elements of all complexes prove to be homogeneous, yet even when this has been recognized, the earlier notion of an antithesis of body and spirit easily slips in again. The philosophical spiritualist is often sensible of the difficulty of imparting the needed solidity to his mind-created world of bodies; the materialist is at a loss when required to endow the world of matter with sensation. The monistic point of view, which reflexion has evolved, is easily clouded by our older and more powerful instinctive notions.

8.

The difficulty referred to is particularly felt when we consider the following case. In the complex $A B C...$, which we have called the world of matter, we find as parts, not only our own body $K L M...$, but also the bodies of other persons (or animals) $K' L' M'...$, $K'' L'' M''...$, to which, by analogy, we imagine other $\alpha' \beta' \gamma'...$, $\alpha'' \beta'' \gamma''...$, annexed, similar to $\alpha \beta \gamma...$So long as we deal with $K' L' M'...$, we find ourselves in a thoroughly familiar province which is at every point accessible to our senses. When, however, we inquire after the sensations or feelings belonging to the body $K' L' M'...$, we no longer find these in the province of sense: we add them in thought. Not only is the domain which we now enter far less familiar to us, but the transition into it is also relatively unsafe. We have the feeling as if we were plunging into an abyss. [8] Persons who adopt this way of thinking only, will never thoroughly rid themselves of that sense of insecurity, which is a very fertile source of illusory problems. But we are not restricted to this course. Let us consider, first, the reciprocal relations of the elements of the complex $A B C...$, without regarding $K L M ...$ (our body). All physical investigations are of this sort. A white ball falls upon a bell; a sound is heard. The ball turns yellow before a sodium lamp, red before a lithium lamp. Here the elements (*A B C*...) appear to be connected only with one another and to be independent of our body (*K L M*...). But if we take santonine, the ball again turns yellow. If we press one eye to the side, we see two balls. If we close our eyes entirely, there is no ball there at all. If we sever the auditory nerve, no sound is heard. The elements *A B C*..., therefore, are not only connected with one another, but also with *K L M*. To this extent, and to this extent *only*, do we call *A B C*...*sensations,* and regard *A B C* as belonging to the ego. In what follows, wherever the reader finds the terms "Sensation," "Sensation-complex," used alongside of or instead of the expressions "element," "complex of elements," it must be borne in mind that it is *only* in the connexion and relation in question, *only* in their functional dependence, that the elements are sensations. In another functional relation they are at the same time physical objects. We only use the additional term "sensations" to describe the elements, because most people are much more familiar with the elements in question as sensations (colors, sounds, pressures, spaces, times, etc.), while according to the popular conception it is particles of mass that are considered as physical elements, to which the elements, in the sense here used, are attached as "properties" or "effects." [9]

In this way, accordingly, we do not find the gap between bodies and sensations above described, between what is without and what is within, between the material world and the spiritual world. [10] All elements *A B C*..., *K L M*..., constitute a *single* coherent mass only, in which, when any one element is disturbed, *all* is put in motion; except that a disturbance in *K L M*...has a more extensive and profound action than one in *A B C*... A magnet in our neighborhood disturbs the particles of iron near it; a falling boulder shakes the earth; but the severing of a nerve sets in motion the *whole* system of elements. Quite involuntarily does this relation of things suggest the picture of a viscous mass, at certain places (as in the ego) more firmly coherent than in others. I have often made use of this image in lectures.

9.

Thus the great gulf between physical and psychological research persists only when we acquiesce in our habitual stereotyped conceptions. A color is a physical object as soon as we consider its dependence, for instance, upon its luminous source, upon other colors, upon temperatures, upon spaces, and so forth. When we consider, however, its dependence upon the retina (the elements *K L M*...), it is a psychological object, a sensation. Not the subject-matter, but the direction of our investigation, is different in the two domains. (Cp. also Chapter Two.)

Both in reasoning from the observation of the bodies of other men or animals, to the sensations which they possess, as well as in investigating the influence of our own body upon our own sensations, we have to complete observed facts by analogy. This is accomplished with much greater ease and certainty, when it relates, say, only to nervous processes, which cannot be fully observed in our own bodies that is, when it is carried out in the more familiar physical domain than when it is extended to the psychical domain, to the sensations and thoughts of other people. Otherwise there is no essential difference.

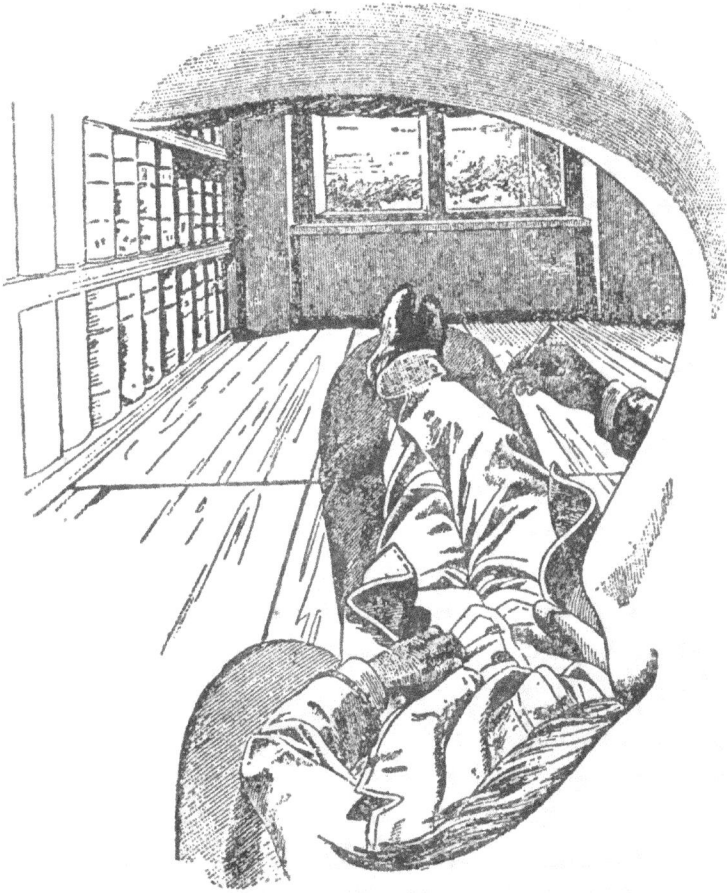

Fig, 1.

10.

The considerations just advanced, expressed as they have been in an abstract form, will gain in strength and vividness if we consider the concrete facts from which they flow. Thus, I lie upon my sofa. . If I close my right eye, the picture represented in the accompanying cut is presented to my left eye. In a frame formed by the ridge of my eyebrow, by my nose, and by my moustache, appears a part of my body, so far as visible, with its environment. [11] My body differs from other human bodies beyond the fact that every intense motor idea is immediately expressed by a movement of it, and that, if it is touched, more striking changes are determined than if other bodies are touched by the circumstance, that it is only seen piecemeal, and, especially, is seen without a head. If I observe an element A within my field of vision, and investigate its connexion with another element B within the same field, I step out of the domain of physics into that of physiology or psychology, provided B, to use the apposite expression of a friend [12] of mine made upon seeing this drawing, passes through my skin. Reflexions like that for the field of vision may be made with regard to the province of touch and the perceptual domains of the other senses. [13]

11.

Reference has already been made to the different character of the groups of elements denoted by $A\ B\ C$...and $\alpha\ \beta\ \gamma$... As a matter of fact, when we see a green tree before us, or remember a green tree, that is, represent a green tree to ourselves, we are perfectly aware of the difference of the two cases. The represented tree has a much less determinate, a much more changeable form; its green is much paler and more evanescent; and, what is of especial note, it plainly appears in a different domain. A movement that we will to execute is never more than a represented movement, and appears in a different domain from that of the executed movement, which always takes place when the image is vivid enough. Now the statement that the elements A and α appear in different domains, means, if we go to the bottom of it, simply this, that these elements are united with different other elements. Thus far, therefore, the fundamental constituents of $A\ B\ C$..., $\alpha\ \beta\ \gamma$...would seem to be *the same* (colors, sounds, spaces, times, motor sensations...), and only the character of their connexion different.

Ordinarily pleasure and pain are regarded as different from sensations. Yet not only tactual sensations, but all other kinds of sensations, may pass gradually into pleasure and pain. Pleasure and pain also may be justly termed sensations. Only they are not so well analysed and so familiar, nor, perhaps, limited to so few organs as the common sensations. In fact, sensations of pleasure and pain, however faint they may be, really constitute an essential part of the content of all so-called emotions. Any additional element that emerges into consciousness when we are under the influence of emotions may be described as more or less diffused and not sharply localized sensations. William James, [14] and after him Theodule Ribot, [15] have investigated the physiological mechanism of the emotions: they hold that what is essential is purposive tendencies of the body to action tendencies which correspond to circumstances and are expressed in the organism. Only a part of these emerges into consciousness. We are sad because we shed tears, and not *vice versa*, says James. And Ribot justly observes that a cause of the backward state of our knowledge of the emotions is that we have always confined our observation to so much of these physiological processes as emerges into consciousness. At the same time he goes too far when he maintains that everything psychical is merely *"surajouté"* to the physical, and that it is only the physical that produces effects. For us this distinction is non-existent.

Thus, perceptions, presentations, volitions, and emotions, in short the whole inner and outer world, are put together, in combinations of varying evanescence and permanence, out of a small number of homogeneous elements. Usually, these elements are called sensations. But as vestiges of a one-sided theory inhere in that term, we prefer to speak simply of elements, as we have already done. The aim of all research is to ascertain the mode of connexion of these elements. [16] If it proves impossible to solve the problem by assuming *one* set of such elements, then more than one will have to be assumed. But for the questions under discussion it would be improper to begin by making complicated assumptions in advance.

12.

That in this complex of elements, which fundamentally is only one, the boundaries of bodies and of the ego do not admit of being established in a manner definite and sufficient for all cases, has already been remarked. To bring together elements that are most intimately connected with pleasure and pain into one ideal mental-economical unity, the ego; this is a task of the highest importance for the intellect working in the service of the pain-avoiding, pleasure-seeking will. The delimitation of the ego, therefore, is instinctively effected, is rendered familiar, and possibly becomes fixed through heredity. Owing to their high practical importance, not only for the individual, but for the entire species, the composites "ego" and "body" instinctively make good their claims, and assert themselves with elementary force. In special cases, however, in which practical ends are not concerned, but where knowledge is an end in itself, the delimitation in question may prove to be insufficient, obstructive, and untenable. [17]

The primary fact is not the ego, but the elements (sensations). What was said earlier as to the term "sensation" must be borne in mind. The elements constitute the I. *I* have the sensation green, signifies that the element green occurs in a given complex of other elements (sensations, memories). When *I* cease to have the sensation green, when *I* die, then the elements no longer occur in the ordinary, familiar association. That is all. Only an ideal mental-economical unity, not a real unity, has ceased to exist. The ego is not a definite, unalterable, sharply bounded unity. None of these attributes are important; for all vary even within the sphere of individual life; in fact their alteration is even sought after by the individual. *Continuity* alone is important. This view accords admirably with the position which Weismann has reached by biological investigations. ("Zur Frage der Unsterblichkeit der Einzelligen," *Biolog. Centralbl.*, Vol. IV., Nos. 21, 22; compare especially pages 654 and 655, where the scission of the individual into two equal halves is spoken of.) But continuity is only a means of preparing and conserving what is contained in the ego. This content, and not the ego, is the principal thing. This content, however, is not confined to the individual. With the exception of some insignificant and valueless personal memories, it remains preserved in others even after the death of the individual. The elements that make up the consciousness of a given individual are firmly connected with one another, but with those of another individual they are only feebly connected, and the connexion is only casually apparent. Contents of consciousness, however, that are of universal significance, break through these limits of the individual, and, attached of course to individuals again, can enjoy a continued existence of an impersonal, superpersonal kind, independently of the personality by means of which they were developed. To contribute to this is the greatest happiness of the artist, the scientist, the inventor, the social reformer, etc.

The ego must be given up. It is partly the perception of this fact, partly the fear of it, that has given rise to the many extravagances of pessimism and optimism, and to numerous religious, ascetic, and philosophical absurdities. In the long run we shall not be able to close our eyes to this simple truth, which is the immediate outcome of psychological analysis. We shall then no longer place so high a value upon the ego, which even during the individual life greatly changes, and which, in sleep or during absorption in some idea, just in our very happiest moments, may be partially or wholly absent. We shall then be willing to renounce individual immortality, [18] and not place more value upon the subsidiary elements than upon the principal ones. In this way we shall arrive at a freer and more enlightened view of life, which will preclude the disregard of other egos and the overestimation of our own. The ethical ideal founded on this view of life will be equally far removed from the ideal of the ascetic, which is not biologically tenable for whoever practises it, and vanishes at once with his disappearance, and from the ideal of an overweening Nietzschean "superman," who cannot, and I hope will not be tolerated by his fellow-men. [19]

If a knowledge of the connexion of the elements (sensations) does not suffice us, and we ask, *Who* possesses this connexion of sensations, *Who* experiences it? then we have succumbed to the old habit of subsuming every element (every sensation) under some unanalysed complex, and we are falling back imperceptibly upon an older, lower, and more limited point of view. It is often pointed out, that a psychical experience which is not the experience of a determinate subject is unthinkable, and it is held that in this way the essential part played by the unity of consciousness has been demonstrated. But the Ego-consciousness can be of many different degrees and composed of a multiplicity of chance memories. One might just as well say that a physical process which does not take place in some environment or other, or at least somewhere in the universe, is unthinkable. In both cases, in order to make a beginning with our investigation, we must be allowed to abstract from the environment, which, as regards its influence, may be very different in different cases, and in special cases may shrink to a minimum. Consider the sensations of the lower animals, to which a subject with definite features can hardly be ascribed. It is out of sensations that the subject is built up, and, once built up, no doubt the subject reacts in turn on the sensations.

The habit of treating the unanalysed ego complex as an indiscernible unity frequently assumes in science remarkable forms. First, the nervous system is separated from the body as the seat of the sensations. In the nervous system again, the brain is selected as the organ best fitted for this end, and finally, to save the supposed psychical unity, a *point* is sought in the brain as the seat of the soul. But such crude conceptions are hardly fit even to foreshadow the roughest outlines of what future research will do for the connexion of the physical and the psychical. The fact that the different organs and parts of the nervous system are physically connected with, and can be readily excited by, one another, is probably at the bottom of the notion of "psychical unity."

I once heard the question seriously discussed, "How the perception of a large tree could find room in the little head of a man?" Now, although this "problem" is no problem, yet it renders us vividly sensible of the absurdity that can be committed by thinking sensations spatially into the brain. When I speak of the sensations of another person, those sensations are, of course, not exhibited in my optical or physical space; they are mentally added, and I conceive them causally, not spatially, attached to the brain observed, or rather, functionally presented. When I speak of my own sensations, these sensations do not exist spatially in my head, but rather my "head" shares with them the same spatial field, as was explained above. (Compare the remarks on Fig. 1, above.) [20]

The unity of consciousness is not an argument in point. Since the apparent antithesis between the real world and the world given through the senses lies entirely in our mode of view, and no actual gulf exists between them, a complicated and variously interconnected content of consciousness is no more difficult to understand than is the complicated interconnexion of the world.

If we regard the ego as a real unity, we become involved in the following dilemma: either we must set over against the ego a world of unknowable entities (which would be quite idle and purposeless), or we must regard the whole world, the egos of other people included, as comprised in our own ego (a proposition to which it is difficult to yield serious assent).

But if we take the ego simply as a practical unity, put together for purposes of provisional survey, or as a more strongly cohering group of elements, less strongly connected with other groups of this kind, questions like those above discussed will not arise, and research will have an unobstructed future.

In his philosophical notes Lichtenberg says: "We become conscious of certain presentations that are not dependent upon us; of others that we at least think are dependent upon us. Where is the border-line? We know only the existence of our sensations, presentations, and thoughts. We should say, *It thinks,* just as we say, *It lightens.* It is going too far to say *cogito,* if we translate *cogito* by *I think.* The assumption, or postulation, of the ego is a mere practical necessity." Though the method by which Lichtenberg arrived at this result is somewhat different from ours, we must nevertheless give our full assent to his conclusion.

13.

Bodies do not produce sensations, but complexes of elements (complexes of sensations) make up bodies. If, to the physicist, bodies appear the real, abiding existences, whilst the "elements" are regarded merely as their evanescent, transitory appearance, the physicist forgets, in the assumption of such a view, that all bodies are but thought-symbols for complexes of elements (complexes of sensations). Here, too, the elements in question form the real, immediate, and ultimate foundation, which it is the task of physiologico-physical research to investigate. By the recognition of this fact, many points of physiology and physics assume more distinct and more economical forms, and many spurious problems are disposed of.

For us, therefore, the world does not consist of mysterious entities, which by their interaction with another, equally mysterious entity, the ego, produce sensations, which alone are accessible. For us, colors, sounds, spaces, times,...are provisionally the ultimate elements, whose given connexion it is our business to investigate. [21] It is precisely in this that the exploration of reality consists. In this investigation we must not allow ourselves to be impeded by such abridgments and delimitations as body, ego, matter, spirit, etc., which have been formed for special, practical purposes and with wholly provisional and limited ends in view. On the contrary, the fittest forms of thought must be created in and by that research itself, just as is done in every special science. In place of the traditional, instinctive ways of thought, a freer, fresher view, conforming to developed experience, and reaching out beyond the requirements of practical life, must be substituted throughout.

14.

Science always has its origin in the adaptation of thought to some definite field of experience. The results of the adaptation are thought-elements, which are able to represent the whole field. The outcome, of course, is different, according to the character and extent of the field. If the field of experience is enlarged, or if several fields heretofore disconnected are united, the traditional, familiar thought-elements no longer suffice for the extended field. In the struggle of acquired habit with the effort after adaptation, problems arise, which disappear when the adaptation is perfected, to make room for others which have arisen meanwhile.

To the physicist, *quâ* physicist, the idea of "body" is productive of a real facilitation of view, and is not the cause of disturbance. So, also, the person with purely practical aims, is materially supported by the idea of the *I* or ego. For, unquestionably, every form of thought that has been designedly or undesignedly constructed for a given purpose, possesses for that purpose a *permanent* value. When, however, physics and psychology meet, the ideas held in the one domain prove to be untenable in the other. From the attempt at mutual adaptation arise the various atomic and monadistic theories which, however, never attain their end. If we regard sensations, in the sense above defined (p. 13), as the elements of the world, the problems referred to appear to be disposed of in all essentials, and the first and most important adaptation to be consequently effected. This fundamental view (without any pretension to being a philosophy for all eternity) can at present be adhered to in all fields of experience; it is consequently the one that accommodates itself with the least expenditure of energy, that is, more economically than any other, to the present temporary collective state of knowledge Furthermore, in the consciousness of its purely economical function, this fundamental view is eminently tolerant. It does not obtrude itself into fields in which the current conceptions are still adequate. It is also ever ready, upon subsequent extensions of the field of experience, to give way before a better conception.

The presentations and conceptions of the average man of the world are formed and dominated, not by the full and pure desire for knowledge as an end in itself, but by the struggle to adapt himself favourably to the conditions of life. Consequently they are less exact, but at the same time also they are preserved from the monstrosities which easily result from a one-sided and impassioned pursuit of a scientific or philosophical point of view. The unprejudiced man of normal psychological development takes the elements which we have called *A B C*...to be spatially contiguous and external to the elements *K L M*..., and he holds this view *immediately,* and not by any process of psychological projection or logical inference or construction; even were such a process to exist, he would certainly not be conscious of it. He sees, then, an "external world" *A B C*... different from his body *K L M*...and existing outside it. As he does not observe at first the dependence of the *A B C's*...on the *K L M's*...(which are always repeating themselves in the same way and consequently receive little attention), but is always dwelling upon the fixed connexion of the *A B C's*...with one another, there appears to him a world of things independent of his Ego. This Ego is formed by the observation of the special properties of the particular thing *K L M*...with which pain, pleasure, feeling, will, etc., are intimately connected. Further, he notices things *K' L' M'*, *K" L" M"*, which behave in a manner perfectly analogous to *K L M*, and whose behaviour he thoroughly understands as soon as he has thought of analogous feelings, sensations, etc., as attached to them in the same way as he observed these feelings, sensations, etc., to be attached to himself. The analogy impelling him to this result is the same as determines him, when he has observed that a wire possesses *all* the properties of a conductor charged with an electric current, except one which has not yet been directly demonstrated, to conclude that the wire possesses this one property as well. Thus, since he does not perceive the sensations of his fellowmen or of animals but only supplies them by analogy, while he infers from the behaviour of his fellow-men that they are in the same position over against himself, he is led to ascribe to the sensations, memories, etc., a particular $A\ B\ C...K\ L\ M...$viz. different nature, always differently conceived according to the degree of civilization he has reached; but this process, as was shown above, is unnecessary, and in science leads into a maze of error, although the falsification is of small significance for practical life.

These factors, determining as they do the intellectual outlook of the plain man, make their appearance alternately in him according to the requirements of practical life for the time being, and persist in a state of nearly stable equilibrium. The scientific conception of the world, however, puts the emphasis now upon one, now upon the other factor, makes sometimes one and sometimes the other its starting-point, and, in its struggle for greater precision, unity and consistency, tries, so far as seems possible, to thrust into the background all but the most indispensable conceptions. In this way dualistic and monistic systems arise.

The plain man is familiar with blindness and deafness, and knows from his everyday experience that the look of things is influenced by his senses; but it never occurs to him to regard the whole world as the creation of his senses. He would find an idealistic system, or such a monstrosity as solipsism, intolerable in practice.

It may easily become a disturbing element in unprejudiced scientific theorizing when a conception which is adapted to a particular and strictly limited purpose is promoted in advance to be the foundation of all investigation. This happens, for example, when all experiences are regarded as "effects" of an external world extending into consciousness. This conception gives us a tangle of metaphysical difficulties which it seems impossible to unravel. But the spectre vanishes at once when we look at the matter as it were in a mathematical light, and make it clear to ourselves that all that is valuable to us is the discovery of *functional relations,* and that what we want to know is merely the dependence of experiences on one another. It then becomes obvious that the reference to unknown fundamental variables which are not given (things-in-themselves) is purely fictitious and superfluous. But even when we allow this fiction, uneconomical though it be, to stand at first, we can still easily distinguish different classes of the mutual dependence of the elements of "the facts of consciousness"; and this alone is important for us.

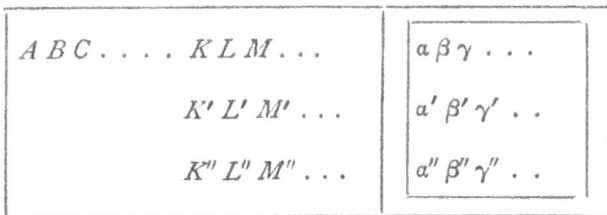

$$
\begin{array}{c|c}
A\ B\ C\ .\ .\ .\ .\ K\ L\ M\ .\ .\ . & \alpha\ \beta\ \gamma\ .\ .\ . \\[4pt]
K'\ L'\ M'\ .\ .\ . & \alpha'\ \beta'\ \gamma'\ .\ . \\[4pt]
K''\ L''\ M''\ .\ .\ . & \alpha''\ \beta''\ \gamma''\ .\ .
\end{array}
$$

The system of the elements is indicated in the above scheme. Within the space surrounded by a single line lie the elements which belong to the sensible world, the elements whose regular connexion and peculiar dependence on one another represent both physical (lifeless) bodies and the bodies of men, animals and plants. All these elements, again, stand in a relation of quite peculiar dependence to certain of the elements $K L M$ the nerves of our body, namely by which the facts of sense-physiology are expressed. The space surrounded by a double line contains the elements belonging to the higher psychic life, memory-images and presentations, including those which we form of the psychic life of our fellow-men. These may be distinguished by accents. These presentations, again, are connected with one another in a different way (association, fancy) from the sensational elements $A B C...K L M$, but it cannot be doubted that they are very closely allied to the latter, and that in the last resort their behaviour is determined by $A B C...K L M$ (the totality of the physical world), and especially by our body and nervous system. The presentations $\alpha' \beta' \gamma'$ of the contents of the consciousness of our fellow-men play for us the part of *intermediate substitutions,* by means of which the behaviour of our fellow-men, the functional relation of $K' L' M'$ to $A B C$ becomes intelligible, in so far as in and for itself (physically) it would remain unexplained.

It is therefore important for us to recognize that in all questions in this connexion, which can be intelligibly asked and which can interest us, everything turns on taking into consideration different *ultimate variables* and different *relations of dependence.* That is the main point. Nothing will be changed in the actual facts or in the functional relations, whether we regard all the data as contents of consciousness, or as partially so, or as completely physical. [22]

The biological task of science is to provide the fully developed human individual with as perfect a. means of orientating himself as possible. No other scientific ideal can be realized, and any other must be meaningless.

The philosophical point of view of the average man if that term may be applied to his naive realism has a claim to the highest consideration. It has arisen in the process of immeasurable time without the intentional assistance of man. It is a product of nature, and is preserved by nature. Everything that philosophy has accomplished though we may admit the biological justification of every advance, nay, of every error is, as compared with it, but an insignificant and ephemeral product of art. The fact is, every thinker, every philosopher, the moment he is forced to abandon his one-sided intellectual occupation by practical necessity, immediately returns to the general point of view of mankind. Professor X., who theoretically believes himself to be a solipsist, is certainly not one in practice when he has to thank a Minister of State for a decoration conferred upon him, or when he lectures to an audience. The Pyrrhonist who is cudgelled in Molière's *Le Manage Forcé,* does not go on saying "Il me semble que vous me battez," but takes his beating as really received.

Nor is it the purpose of these "introductory remarks" to discredit the standpoint of the plain man. The task which we have set ourselves is simply to show why and for what purpose we hold that standpoint during most of our lives, and why and for what purpose we are provisionally obliged to abandon it. No point of view has absolute, permanent validity. Each has importance only for some given end.

[1] Once, when a young man, I noticed in the street the profile of a face that was very displeasing and repulsive to me. I was not a little taken aback when a moment afterwards I found that it was my own face which, in passing by a shop where mirrors were sold, I had perceived reflected from two mirrors that were inclined at the proper angle to each other.

Not long ago, after a trying railway journey by night, when I was very tired, I got into an omnibus, just as another man appeared at the other end. "What a shabby pedagogue that is, that has just entered," thought I. It was myself: opposite me hung a large mirror. The physiognomy of my class, accordingly, was better known to me than my own.

[2] Cp. Hume, *Treatise on Human Nature,* Vol. I. part iv., p. 6; Gruithuisen, *Beiträge zur Physiognosie und Eautognosie,* Munich, 1812, pp. 37-58.

[3] Not to be taken in the metaphysical sense.

[4] If this process be regarded as an abstraction, the elements, as we shall see, do not thereby lose anything of their importance. Cp. the subsequent discussion of concepts in Chapter Fourteen.

[5] Cp. W. Schuppe's polemic against Ueberweg, printed in Brasch's *Welt-und Lebensanschauung Ueber-wegs,* Leipzig, 1889; F.J. Schmidt, *Das Aergernis der Philosophie: eine Kantstudie,* Berlin, 1897.

[6] A long time ago (in the *Vierteljahrsschrift für Psychiatrie,* Leipzig and Neuwied, 1868, art. "Ueber die Abhängigkeit der Netzhautstellen von einander") I enunciated this thought as follows: The expression "sense-illusion" proves that we are not yet fully conscious, or at least have not yet deemed it necessary to incorporate the fact into our ordinary language, *that the senses represent things neither wrongly nor correctly.* All that can be truly said of the sense-organs is, that, *under different circumstances they produce different sensations and perceptions.* As these "circumstances," now, are extremely various in character, being partly external (inherent in the objects), partly internal (inherent in the sensory organs), and partly interior (having their activity in the central organs), it can sometimes appear, when we only notice the external circumstances, as if the organ acted differently under the same conditions. And it is customary to call the unusual effects, deceptions or illusions.

[7] When I say that the table, the tree, and so forth, are my sensations, the statement, as contrasted with the mode of representation of the ordinary man, involves a real extension of my ego. On the emotional side also such extensions occur, as in the case of the virtuoso, who possesses as perfect a mastery of his instrument as he does of his own body; or in the case of the skilful orator, on whom the eyes of the audience are all converged, and who is controlling the thoughts of all; or in that of the able politician who is deftly guiding his party;

and so on. In conditions of depression, on the other hand, such as nervous people often endure, the ego contracts and shrinks. A wall seems to separate it from the world.

[8] When I first came to Vienna from the country, as a boy of four or five years, and was taken by my father upon the walls of the city's fortifications, I was very much surprised to see people below in the moat, and could not understand how, from my point of view, they could have got there; for the thought of another way of descent never occurred to me. I remarked the same astonishment, once afterwards, in the case of a three-year-old boy of my own, while walking on the walls of Prague. I recall this feeling every time I occupy myself with the reflexion of the text, and I frankly confess that this accidental experience of mine helped to confirm my opinion upon this point, which I have now long held. Our habit of always following the same path, whether materially or psychically, tends greatly to confuse our field of survey. A child, on the piercing of the wall of a house in which he has long dwelt, may experience a veritable enlargement of his worldview, and in the same manner a slight scientific hint may often afford great enlightenment.

[9] A treatment of this fundamental point, identical in essentials, but cast in a form which will be perhaps more acceptable to scientists, will be found in my *Erkenntnis und Irrtum*, Leipzig, 1905 (2nd edition, Leipzig, 1906.)

[10] Compare my *Grundlinien der Lehre von den Bewegungsempfindungen*, Leipzig, Engelmann, 1875, p. 54. I there, for the first time, stated my view shortly, but definitely, in these words: "Appearance may be subdivided into elements, which, in so far as they are connected with certain processes of our bodies, and can be regarded as conditioned by these processes, we call sensations."

[11] A discussion of the binocular field of vision, with its peculiar stereoscopic features, is omitted here, for although familiar to all, it is not as easy to describe, and cannot be represented by a single plane drawing.

[12] J. Popper of Vienna.

[13] It was about 1870 that the idea of this drawing was suggested to me by an amusing chance. A certain Mr L., now long dead, whose many eccentricities were redeemed by his truly amiable character, compelled me to read one of C. F. Krause's writings, in which the following occurs:-

"Problem: To carry out the self-inspection of the Ego.

Solution: It is carried out immediately."

In order to illustrate in a humorous manner this philosophical "much ado about nothing," and at the same time to show how the self-inspection of the Ego could be really "carried out," I embarked on the above drawing. Mr. L.'s society was most instructive and stimulating to me, owing to the naivety with which he gave utterance to philosophical notions that are apt to be carefully passed over in silence or involved in obscurity.

[14] W. James, *Psychology*, New York, 1890, II., p. 442.

[15] Th. Ribot, *La psychologie des sentiments*, 1896. (English translation, *The Psychology of the Emotions*, 1897.)

[16] Compare the note at the conclusion of my treatise, *Die Geschichte und die Wurzel des Satzes der Erhaltung der Arbeit*, Prague, Calve, 1872. (*History and Root of the Principle of the Conservation of Energy*. Translated and annotated by P. E. B. Jourdain, Chicago, Open Court Publishing Co., 1911.)

[17] Similarly, class-consciousness, class-prejudice, the feeling of nationality, and even the narrowest-minded local patriotism may have a high importance, *for certain purposes*. But such attitudes will not be shared by the broad-minded investigator, at least not in moments of research. All such egoistic views are adequate only for practical purposes. Of course, even the investigator may succumb to habit. Trifling pedantries and nonsensical discussions; the cunning appropriation of others' thoughts, with perfidious silence as to the sources; when the word of recognition must be given, the difficulty of swallowing one's defeat, and the too common eagerness at the same time to set the opponent's achievement in a false light: all this abundantly shows that the scientist and scholar have also the battle of existence to fight, that the ways even of science still lead to the mouth, and that the pure impulse towards knowledge is still an ideal in our present social conditions.

[18] In wishing to preserve our personal memories beyond death, we are behaving like the astute Eskimo, who refused with thanks the gift of immortality without his seals and walruses.

[19] However far the distance is from theoretical understanding to practical conduct, still the latter cannot in the long run resist the former.

[20] As early as the writings of Johannes Müller, we can already find a tendency towards views of this kind, although his metaphysical bias prevents him from carrying them to their logical conclusion. But Hering (Hermann's *Handbuch der Physiologie*, Vol. III., p. 345) has the following characteristic passage: "The material of which visual objects consists is the visual sensations. The setting sun, as a visual object, is a flat, circular disk, which consists of yellowish-red color, that is to say of a visual sensation. We may therefore describe it directly as a circular, yellowish-red sensation. This sensation we have in the very place where the sun appears to us." I must confess that, so far as the experiments go which I have had occasion to make in conversation, most people, who have not come to close quarters with these questions by serious thinking, will pronounce this way of looking at the matter to be mere hair-splitting. Of course, what is chiefly responsible for their indignation is the common confusion between sensible and conceptual space. But anyone who takes his stand as I do on the *economic* function of science, according to which nothing is important except what can be observed or is a datum for us, and everything hypothetical, metaphysical and superfluous, is to be eliminated, must reach the same conclusion. I think that a similar standpoint is to be ascribed to Avenarius, for in his *Der menschliche Weltbegriff*, p. 76, the following passages occur: "The brain is not the dwelling-place, seat or producer of thought; it is not the instrument or organ, it is not the vehicle or substratum, etc., of thought." "Thought is not an indweller or command-giver, it is not a second half or aspect, etc., nor is it a product; it is not even a physiological function of the brain, nor is it a state of the brain at all. "I am not able or willing to subscribe to all that Avenarius says or to any interpretation of what he says, but still his conception seems to me to approximate very nearly to my own. The method which he terms, "The exclusion of introjection," is only a particular form of the elimination of the metaphysical.

[21] I have always felt it as a stroke of special good fortune, that early in life, at about the age of fifteen, I lighted, in the library of my father, on a copy of Kant's

Prolegomena to any Future Metaphysics. The book made at the time a powerful and ineffaceable impression upon me, the like of which I never afterwards experienced in any of my philosophical reading. Some two or three years later the superfluity of the role played by "the thing in itself" abruptly dawned upon me. On a bright summer day in the open air, the world with my ego suddenly appeared to me as one coherent mass of sensations, only more strongly coherent in the ego. Although the actual working out of this thought did not occur until a later period, yet this moment was decisive for my whole view. I had still to struggle long and hard before I was able to retain the new conception in my special subject. With the valuable parts of physical theories we necessarily absorb a good dose of false metaphysics, which it is very difficult to sift out from what deserves to be preserved, especially when those theories have become very familiar to us. At times, too, the traditional, instinctive views would arise with great power and place impediments in my way. Only by alternate studies in physics and in the physiology of the senses, and by historico-physical investigations (since about 1863), and after having endeavored in vain to settle the conflict by a physico-psychological monadology (in my lectures on psycho-physics, in the *Zeitschrift für praktische Heilkunde,* Vienna, 1863, p. 364), have I attained to any considerable stability in my views. I make no pretensions to the title of philosopher. I only seek to adopt in physics a point of view that need not be changed the moment our glance is carried over into the domain of another science; for, ultimately, all must form one whole. The molecular physics of to-day certainly does not meet this requirement. What I say I have probably not been the *first* to say. I also do not wish to offer this exposition of mine as a special achievement. It is rather my belief that every one will be led to a similar view, who makes a careful survey of any extensive body of knowledge. Avenarius, with whose works I became acquainted in 1883, approaches my point of view (*Philosophie als Denken der Welt nach dem Princip des kleinsten Kraftmasses,* 1876). Also Hering, in his paper on *Memory* (*Almanach der Wiener Akademie,* 1870, p. 258; English translation, O. C. Pub. Co., Chicago, 4th edition, enlarged, 1913), and J. Popper in his beautiful book, *Das Rechte zu leben und die Pflicht zu sterben* (Leipzig, 1878, p. 62), have advanced allied thoughts. Compare also my paper *Ueber die ökonomische Natur der physikalischen Forschung* (*Almanach der Wiener Akademie,* 1882, p. 179, note; English translation in my *Popular Scientific Lectures,* Chicago, 1894). Finally let me also refer here to the introduction to W. Preyer's *Reine Empfindungslehre,* to Riehl's *Freiburger Antrittsrede,* p. 40, and to R. Wahle's *Gehirn und Bewusstsein,* 1884. My views were indicated briefly in 1872 and 1875, and not expounded at length until 1882 and 1883. I should probably have much additional matter to cite as more or less allied to this line of thought, if my knowledge of the literature were more extensive.

[22] Cf. J. Petzoldt's excellent paper "Solipsismus auf praktischem Gebiet" (*Vierteljahrsschrift für wissentschaftliche Philosophie,* XXXV., 3, p. 339) Schuppe, "Der Solipsismus" (*Zeitschr für immanente Philosophie,* Vol. III., p. 327).

II. On Preconceived Opinions

1.

THE physicist has frequent occasion to observe how greatly our knowledge of some field of research may be hampered, when, instead of the unprejudiced investigation of that field for its own sake, views are transferred to it which have been formed in some other department of knowledge. Far more serious is the confusion which arises from such transference of preconceived opinions from the field of physics to that of psychology. Let us illustrate this by a few examples.

A physicist observes the inverted image on the retina of an excised eye, and puts to himself very naturally the question, How does a point situated low down in space come to be reflected on the upper part of the retina? He answers this question by the aid of studies in dioptrics. If, now, this question, which is perfectly legitimate in the province of physics, be transferred to the domain of psychology, only obscurity will be produced. The question why we see inverted retinal images the right way up, has no meaning as a psychological problem. The light-sensations of the separate spots of the retina are connected with space-sensations from the very outset, and we give the name "above" to those positions in space that correspond to the parts of the retina situated lower down. To the subject having the sensation such a question cannot present itself.

It is the same with the well-known theory of external projection. The problem of the physicist is, to find the luminous object-point corresponding to a point on the retinal image, by prolonging the ray that passes through the point on the image and the centre of the eye. For the subject having the sensation this problem does not exist, as the light sensations are connected from the beginning with determinate space-sensations. The entire theory of the psychological origin of the external world by the projection of sensations outwards is founded solely on a mistaken application of physical points of view. Our sensations of sight and touch are bound up with various different sensations of space; that is to say, they exist by the side of one another and outside one another, exist, in other words, in a spatial field, in which our body fills but a part. The table, the house, the tree, lie thus self-evidently outside my body. A projection-problem never presents itself, is neither consciously nor unconsciously solved.

A physicist (Mariotte) discovers that a certain spot on the retina is blind. The physicist is accustomed to correlate with every spatial point a point on the retinal image, and with every point on the image a sensation. Hence the question arises, What do we see at the points corresponding to the blind spot, and how is the gap filled out? When the illegitimate form of putting the question in physical terms is eliminated from the psychological inquiry, it will be found that no problem exists at all here. We see nothing at the blind spots, the gap in the image is not filled out, or rather, the gap is not felt, for the simple reason that a defect of light-sensation can no more be noticed at a spot blind from the beginning than the blindness, say, of the skin of the back can cause a gap in the visual field.

I have intentionally chosen simple and obvious examples, as they can best make clear what unnecessary confusion is caused by the thoughtless transference of a conception or mode of thought, which is valid and serviceable in one field, into another quite different field.

In the work of a celebrated German ethnographer I recently read the following sentence: "This tribe has become deeply degraded through the practice of cannibalism." By its side lay the book of an English inquirer who deals with the same subject. The latter simply puts the question why certain South-Sea Islanders are cannibals, finds out in the course of his inquiries that our own ancestors also were once cannibals, and comes to understand the position the Hindus take in the matter a point of view that occurred once to my five-year-old boy who while eating a piece of meat stopped, suddenly shocked, and cried out, "*We* are cannibals to the animals!" "Thou shalt not eat human beings" is a very praiseworthy maxim; but in the mouth of the ethnographer it destroys that mild and sublime glow of freedom from prepossession by which we delight to recognize the true inquirer. But a step further, and we shall say, "Man *must* not be descended from monkeys," "The earth *ought* not to rotate," "Matter *ought* not to fill space continuously," "Energy must be constant," and so on. I believe that our procedure differs from that just characterized only in degree and not in kind, when we claim absolute validity for views reached in the field of physics, and transfer them to the field of psychology without having first tested their applicability. In such cases we succumb to dogma, if not, like our scholastic forefathers, to dogma forced upon us from without, yet to that which we have created ourselves. And what result of research is there that could not become a dogma by long habit? The very skill which we have acquired to deal with constantly recurring intellectual situations deprives us of the freshness and open-mindedness which we so greatly need in new situations.

After these general remarks, I may perhaps be able to explain my position with regard to the dualism of the physical and the psychical. This dualism is to my mind artificial and unnecessary.

2.

In the investigation of purely physical processes we generally employ concepts of so abstract a character that as a rule we think only cursorily, or not at all, of the sensations (elements) that lie at their base. For example, when I ascertain the fact that an electric current having the intensity of 1 ampere develops 10½ cubic centimetres of oxyhydrogen gas at 0°C. and 760 mm. mercury-pressure in a minute, I am readily disposed to attribute to the objects defined a reality wholly independent of my sensations. But I am obliged, in order to arrive at what I have defined, to conduct the current, for the existence of which my sensations are my only warrant, through a circular wire having a definite radius, so that the current, the intensity of terrestrial magnetism being given, shall turn the magnetic needle a certain angular distance out of the meridian. The determination of the magnetic intensity, of the volume of the oxyhydrogen gas, etc., is no less intricate. The whole statement is based upon an almost unending series of sensations, particularly if we take into consideration the adjustment of the apparatus, which must precede the actual experiment. Now it can easily happen to the physicist who does not study the psychology of his operations, that he does not (to reverse a well-known saying) see the trees for the wood, that he overlooks the sensory elements at the foundation of his work. Now I maintain that every physical concept means nothing but a certain definite kind of connexion of the sensory elements which I have denoted by *A B C*... These elements - elements in the sense that no further resolution has as yet been made of them - are the simplest materials out of which the physical, and also the psychological, world is built up.

Physiological research also may be of a purely physical character. I can follow the course of a physical process as it propagates itself through a sensitive nerve to the central organ; I can thence trace it by various paths to the muscles, whose contraction produces new physical changes in the environment. In so doing I am precluded from thinking of any sensation felt by the man or animal under observation; what I am investigating is a purely physical object. Very much is lacking, it is true, to our complete comprehension of the details of this process, and the assurance that everything depends on "the motion of molecules" can neither console me nor deceive me with respect to my ignorance.

Long prior to the development of a scientific psychology people had nevertheless perceived that the behaviour of an animal under physical influences can be predicted with much greater accuracy, *i.e.*, can be better understood, if we attribute to the animal sensations and memories like our own. To that which I observe, to my sensations, I have to supply mentally the sensations of the animal, which are not to be found in the field of my own sensations. This antithesis appears even more abrupt to the scientific inquirer who is investigating a nervous process by the aid of colorless abstract concepts, and is required, for example, to add mentally to that process the sensation green. This last may actually appear as something entirely new and strange, and we ask ourselves how it is that this miraculous thing is produced from chemical processes, electrical currents, and the like.

3.

Psychological analysis has taught us that this surprise is unjustifiable, since the physicist is always operating with sensations. The same analysis also shows us that the process of mentally supplementing complexes of sensations according to analogy by means of elements which at the moment are not being observed, or by elements which cannot possibly be observed, is one which is daily practised by the physicist; as, for example, when he imagines the moon a tangible, inert, heavy mass. The totally strange character of the intellectual situation above described is therefore an illusion.

There is also another consideration, - a consideration confined to my own sensory sphere, - which serves to dispel the illusion. Before me lies the leaf of a plant.

The green (A) of the leaf is connected with a certain optical sensation of space (B), with a sensation of touch (C), and with the visibility of the sun or the lamp (D). If the yellow (E) of a sodium flame takes the place of the sun, the green (A) will pass into brown (F). If the chlorophyl granules be removed by alcohol, an operation which can be represented, like the preceding one, by elements, - the green (A) will pass into white (G). All these observations are physical observations. But the green (A) is also connected with a certain process of my retina. There is nothing to prevent me in principle from investigating this process in my own eye in exactly the same manner as in the previous cases, and from reducing it to its elements $X\,Y\,Z$...If there are difficulties in doing this for my own eye, it can be done with some one else's eye, and the gap filled out by analogy, exactly as in other physical investigations. Now in its dependence upon $B\,C\,D$..., A is a physical element, in its dependence on $X\,Y\,Z$...it is a sensation, and can also be considered as a psychical element. The green (A), however, is not altered at all in itself, whether we direct our attention to the one or to the other form of dependence. I see, therefore, no opposition of physical and psychical, but simple identity as regards these elements. In the sensory sphere of my consciousness everything is at once physical and psychical.

4.

The obscurity of this intellectual situation has, I take it, arisen solely from the transference of a physical prepossession to the field of psychology. The physicist says: I find everywhere bodies and the motions of bodies only, no sensations; sensations, therefore, must be something entirely different from the physical objects I deal with. The psychologist accepts the second portion of this declaration. For him, as is proper, sensations are the primary *data;* but to these there corresponds a mysterious physical something which, conformably with the prepossession, must be quite different from sensations. But what is it that is the really mysterious thing? Is it the Physis or the Psyche? Or is it perhaps both? It would almost appear so, as it is now the one and now the other that appears unattainable and involved in impenetrable obscurity. Or are we here being led round in a circle by some evil spirit?

I believe that the latter is the case. For me the elements $A\,B\,C$...are immediately and indubitably given, and for me they can never afterwards be volatilized away by considerations which ultimately are always based on their existence.

The task of specialized investigation in the sensory physico-psychical sphere, which has not been made superfluous by this general survey, is to ascertain the peculiar method of combination of the *A B C s*. This may be expressed symbolically by saying that it is the object of special research to find equations of the form $f(A B C...) = o$ (zero).

III. My Relation to Richard Avenarius and Other Thinkers

1.

I HAVE already alluded to points at which the views here advocated are in touch with those of various philosophers and philosophically inclined scientists. A full enumeration of these points of contact would require me to begin with Spinoza. That my starting-point is not essentially different from Hume's is of course obvious. I differ from Comte in holding that the psychological facts are, as sources of knowledge, at least as important as the physical facts. My position, moreover, borders closely on that of the representatives of the philosophy of immanence. This is especially true in the case of Schuppe, with whose writings I became acquainted in 1902; his *Outline of Theory of Knowledge and Logic,* a work which is packed with thought and which can be read without a special dictionary, struck a particularly sympathetic chord in me. In this book I have found scarcely anything to which, perhaps with some small modification, I cannot yield a hearty assent. To be sure, his conception of the Ego constitutes a point of difference between us; but not a point on which it would be hopeless to reach an understanding. As to the views of Avenarius, the affinity between them and my own is as great as can possibly be imagined where two writers have undergone a different process of development, work in different fields, and are completely independent of one another. The agreement is somewhat obscured by the great difference of form. Avenarius presents us with a scheme, exhaustive indeed but highly generalized, which is made more difficult to grasp by a strange and unfamiliar terminology. For constructions of this kind I had neither occasion nor vocation, neither inclination nor talent; I am a scientist and not a philosopher. What I aimed at was merely to attain a safe and clear philosophical standpoint, whence practicable paths, shrouded in no metaphysical clouds, might be seen leading not only into the field of physics but also into that of psycho-physiology. With the attainment of this, my battle was won. Although my theory is the fruit of long years of meditation from earliest youth upwards, yet in its brevity it has the form of a mere *aperçu,* nor shall I be offended if it is regarded in that light. I willingly admit that in my distaste for an artificial terminology I have perhaps fallen into the opposite extreme to that of Avenarius. While Avenarius is often not to be understood, or only to be understood after much study, my words have often enough been misunderstood. One acute critic considers that many of my results are results which I *ought* not to have reached (he is therefore in a position to save himself the trouble of investigation, since he already knows the results to which investigation *ought* to lead), and proceeds to reproach me with being difficult to place, since I use ordinary language, and it is consequently impossible to see to what "system" I adhere. Thus, first and foremost, you have to choose a system; then within the walls of that system you may think and speak. In this way all kinds of current popular views have been comfortably read into my words; I have been accused of idealism, Berkeleyanism, even of materialism, and of other "-isms," of all of which I believe myself to be innocent.

The fact is that each of the two methods of exposition has its advantages and disadvantages. But the difference of form has had a prejudicial effect also on the mutual understanding between Avenarius and myself. I recognized the affinity between our views at a very early stage, and expressed my conviction of it in the *Mechanik* (1883) and in the first edition of this book (1886), although at that time I was only able to refer to one of Avenarius' minor works (*Denken der Welt nach dem Prinzip des kleinsten Kraftmasses*), which had appeared in 1876 and had accidentally come in my way shortly before the publication of my *Mechanik*. It was only in 1888, 1891, and 1894, by means of Avenarius' publications, *Kritik der reinen Erfahrung, Der menschliche Weltbegriff,* and his psychological articles in the *Vierteljahrsschrift,* that the similarity of our tendencies was fully revealed to me. As to the first of these works, however, its somewhat hypermetaphorical terminology prevented me from tasting the full rapture of agreement. It is asking rather much of an elderly man that to the labour of learning the languages of the nations he should add that of learning the language of an individual. It was consequently reserved for the younger generation to turn the work of Avenarius to good use: in this connexion I am glad to be able to refer to the writings of H. Cornelius, C. Hauptmann and J. Petzoldt, which are in a fair way to bring to light and to develop further the real value of Avenarius' work. Avenarius, too, acknowledged on his side the affinity between us, and noticed it in the books that appeared from 1888 to 1895. Yet with him, too, the conviction of a profounder coincidence of view between us seems, as I am forced to infer from remarks made by him in the past to third persons, only to have been developed gradually. The man himself I have never known personally. Unmistakable efforts are being made to minimize his importance, but, in spite of that, acquaintance with his works is, I am glad to say, on the increase,

2.

I should now like to indicate more particularly those points of agreement between us to which I attach importance. The economy of thought, the economical representation of the actual, this was indicated by me, in summary fashion first in 1871 and 1872, as being the essential task of science, and in

1882 and 1883 I gave considerably enlarged expositions of this idea. As I have shown elsewhere, this conception, which implicitly contains and anticipates Kirchhoff's notion of "perfectly simple description" (1874), was by no means quite new; it can be traced back to Adam Smith, and, as P. Volkmann holds, in its beginnings even to Newton. We find the same conception again, with the exception of one feature that does not come out clearly, fully developed in Avenarius (1876).

A broad foundation is laid for the theory in question, and light is shed upon it from new sides, if, in conformity with the stimulus given by Darwinism, we conceive of all psychical life including science as biological appearance, and if we apply to the theory the Darwinian conceptions of struggle for existence, of development, and of selection. The theory is inseparable from the hypothesis that each and every psychical entity is physically founded and determined. Now, in his *Kritik der reinen Erfahrung,* Avenarius tries to show in detail that all theoretical and practical activity is determined by change in the central nervous system. In doing this he is merely basing himself on the very general assumption that the central organ is subject to an impulse of self-preservation, a tendency to maintain its equilibrium, not only as a whole, but also in its parts. This agrees very well with the conceptions developed by Hering as to the behaviour of living substance. In holding these views, Avenarius is brought into close contact with modern positive research, particularly in physiology. In my writings too, opinions of a corresponding nature, briefly indeed, but definitely expressed, appeared as long ago as 1863, and in 1883 I expounded these opinions at greater length, though without developing a complete system, such as we find in Avenarius.

But it is to our agreement in the conception of the relation between the physical and the psychical that I attach the greatest importance. For me this is the main point at issue. It was by means of his psychological articles that I first became convinced of this coincidence between Avenarius and myself. In order to be sure of making no mistake, I addressed an inquiry on the subject to Dr. Rudolf Wlassak, who I knew would be intimately acquainted with Avenarius' standpoint, thanks to his association with him for many years. His reply was as follows:

"The conception of the relation of 'the physical' to 'the psychical' is identical in Avenarius and Mach. Both come to the conclusion that the difference between the physical and the psychical consists solely in the difference of the relations of dependence, which on the one hand are the objects dealt with by Physics (in the widest sense of the word), and on the other are the objects dealt with by psychology. If I investigate the dependence of one constituent (*A*) of an environment on another constituent (*B*) of an environment, I am studying physics; if I inquire to what extent A is changed by a change in the sense-organs or the central nervous system, I am studying psychology. Avenarius has accordingly proposed to abolish the terms 'physical' and 'psychical' and in future only to speak of physical and psychological dependencies ("Observations," *Vierteljahrsschrift,* xix., p. 18). In Mach's work the same view occurs, except that (?) the untenability of the old conception of the psychical, and consequently of the proper task of psychology, is not demonstrated.

"This task is performed by the exposure of 'Introjection,' or rather of the fallacy in formal logic which underlies introjection. Avenarius starts from the fact that naive realism, 'the natural view of the world,' stands at the beginning of all philosophizing. Within the limits of this natural view of the world it is possible for a relative delimitation of the complex 'self and the complex 'environment' to be carried out without necessarily involving the 'dualism' of 'body' and 'soul,' since, from the standpoint of naive realism, the constituents that belong to the 'self,' to one's own body, are through and through comparable with the constituents of the environment.

Even when the preliminary survey has advanced to the formation of concepts of substance (Mach, *Analysts of Sensations,* p. 5), this does not mean that a complete and essential difference between body and soul is given. The final splitting up of the world, originally conceived by naive realism as a unity, is really occasioned, according to Avenarius, by the interpretation of the utterances of our fellow-men. As long as I say, 'the tree does not exist for me alone; it also, as their utterances permit me to assume, exists for my fellow-men in the same way as it exists for me,' I am in no wise overstepping the analogy which formal logic allows between me and my fellow-men. But I am overstepping this analogy, if I say that the tree exists as 'a copy,' ('a sensation' or 'a presentation' in my fellowmen, if I *introduce* or *introject* the tree; since I am then assuming something for my fellow-men which I cannot discover in my own experience, which always shews me the constituents of my environment as standing in a definite relation to my body, and never as being in my consciousness or the like. Inasmuch as introjection is a way of passing beyond experience, every attempt to bring it into harmony with the facts of experience must become an inexhaustible source of pseudo-problems. This is most clearly seen in the different forms which, in the course of the history of philosophy, introjection has assumed. The oldest and crudest theories of perception exhibited it in its crudest and simplest form; they supposed that copies were detached from objects, and that these copies penetrated within the body. Now in so far as it is recognized that the constituents of the environment are *not* present inside the body in the same way as they are present outside it, to that extent they are bound, the moment they are inside it, to become something essentially different from the environment. The root of dualism lies in the extension of introjection, in the attempt to bring it into harmony with the complex of the environment.

"It may be doubted whether Avenarius' account of the motives for introjection is satisfactory in all cases. He holds that introjection is always connected with the explanation of the 'perceptions' of one's fellow-men. But it might well be urged that the fact that one and the same constituent of the environment is at one time given in sensations as a 'thing,' and at another given as a memory, can be sufficient motive for conceiving this constituent as being present twice over, namely, first 'materially' in the environment, and secondly in my 'consciousness,' in my 'soul.' Another point to be considered is, whether dream-experiencies cannot equally constitute, at a primitive stage of culture, an independent motive for dualism. [1] Avenarius, indeed, represents introjection as the presupposition of the dualist's interpretation of dream-experiences, but without adducing conclusive reasons. But it is not justifiable to regard prehistoric animism as the root of dualism, if by 'animism' we understand merely the hypothesis that all the lifeless constituents of our environment are beings like ourselves. As long as deeper-lying physiological reasons do not make it impossible, the 'natural view of the world' can also provide a foundation for the hypothesis that, in the case of the tree, for example, the constituents of its environment exist for it in the same sense as they do for human beings. In other words: Anyone who held the view of the psychical common to Avenarius and Mach, might, if he were entirely ignorant of physiology, suppose that a tree or a stone touches and sees its environment. In that case he would still not be a dualist. He only becomes a dualist when, in order to explain this touching and seeing on the part of the tree or stone, he assumes that the constituents of the environment, which the tree and stone taste and see, are present over again in the tree as its 'sensations' or its 'consciousness.' It is only then that the world is duplicated by division into a spiritual part and a material part.

"The discovery of the illegitimacy of introjection throws light in two directions. On the one hand it is illuminating on the side of theory of knowledge. All problems connected with the relation of our 'sensations,' 'presentations' and 'contents of consciousness' to the material things, of which the above-mentioned products of introjection are supposed to be the 'copies,' 'signs,' and so forth, are seen to be merely illusory. Instances of such pseudo-problems are the problems as to projection which we meet in theories of space, the exteriorization of the space-sensations, etc.

"On the other hand, the elimination of introjection implies that all psychology which is not physiological is illegitimate. When I have recognized that the 'contents of consciousness,' the 'psychic processes,' which accompany changes in the nervous system, are nothing more than constituents of my environment which I have introduced into my fellowmen and ultimately also into myself, it is impossible for me to look for anything in the nervous system except physiological processes. All special psychical causality disappears; all those problems disappear which are connected with the question whether the intervention of psychical forces in the physiological processes of the brain is compatible with the principle of the conservation of energy. [2]

"When such a phrase is used as 'the continued existence of presentations without their being in consciousness' (Mach, *Wärmelehre* p. 441), this is, strictly speaking, only legitimate as an abbreviated expression for particular processes of the central nervous system, and in any case it savours strongly of dualistic conceptions."

The remaining difference between the way in which Avenarius puts his views and the way in which I put mine can be reduced to elements which are easily grasped. In the first place, it was not my intention to give a complete exposition of the development of my point of view out of the preceding phrases of philosophical reflection about the world. In the second place, Avenarius starts from a realistic phase; I, on the other hand, started from an idealist phase (p. 30, above, footnote), such as I actually went through in my early youth. In that way I might easily have talked, for instance, about the elimination of "extrajection" (pp. 6, 12-22, 29-35, 43, above). In the third place, it is not necessary to attribute so important a role to the interpretation of our fellow-men by means of introjection in the bad sense, until after the new standpoint has been reached; and then it is also not necessary to exclude this introjection again. A solitary thinker might reach the new standpoint, and even he, as Wlassak observes, might have to rise superior to dualistic tendencies. But this standpoint once reached, and the varying character of the dependence . of the elements once recognized as the essential point, the question whether we start from a phase of realism or of idealism is of no greater importance for us than a change in the fundamental variables of his equations is for the mathematician or physicist.

What Avenarius puts forward, and consequently what I also put forward, appears to me to contain scarcely anything that is not self-evident - self-evident at least for every man who has shaken himself free from the pressure of "the legacy of wild philosophy," as Tylor calls it. Science has always required such self-evident propositions as a safe foundation upon which to build. When I see the paths pursued by different philosophical thinkers converging, and especially when I contemplate the close coincidence between general philosophical views and the views of scientific specialists, I think that I am justified in detecting here a hopeful presage of the mutual accommodation of the sciences to one another.

[1] According to Tylor, they are in fact one of the strongest motives (Mach).

[2] I cannot refrain from here expressing my surprise that the principle of the conservation of energy has so often been dragged in in connexion with the question whether there is a special psychical agent. On the assumption that energy is constant, the course of physical processes is *limited,* but not necessarily determined with perfect *uniqueness*. That the principle of conservation of energy is satisfied in all physiological cases, merely tells us that the soul neither uses up work nor performs it. For all that, the soul may still be a partly determinant factor. When the philosopher asks a question which has reference to this case, he usually misses the point of the principle of the conservation of energy, and the stock reply of the physicist has no intelligible meaning in relation to a case so far removed from the scope of his ideas. Cp. the references to a similar discussion in Höfler's *Psychologie,* 1897, pp. 58 sqq., note. Apart from the above considera-

tions, the assumption of a special psychical agent appears to me to be a presupposition which is unfortunate and can only do harm by making investigation difficult; it is moreover unnecessary and improbable (Mach).

IV. The Chief Points of View for the Investigation of the Senses

1.

IN order to get our bearings, we will now try to obtain, from the standpoint we have reached, a broad view of the special problems that will engage our attention.

When once the inquiring intellect has formed, through adaptation, the habit of connecting two things, A and B, in thought, it tries to retain this habit as far as possible, even where the circumstances are slightly altered. Wherever A appears, B is added in thought. The principle thus expressed, which has its root in an effort for economy, and is particularly noticeable in the work of the great investigators, may be termed the *principle of continuity*.

Every actually observed variation in the connexion of A and B which is sufficiently large to be noticed makes itself felt as a disturbance of the abovementioned habit, and continues to do so until the habit is sufficiently modified to prevent the disturbance being felt. Suppose, for instance, that we have become accustomed to seeing light deflected when it impinges on the boundary between air and glass. But these deflections vary noticeably in different cases, and the habit formed by observing some cases cannot be transferred undisturbed to new cases, until we are able to associate with every particular angle of incidence (A) a particular angle of refraction (B), which we are able to do by discovering the so-called law of refraction, and by making ourselves familiar with the rules contained in that law. Thus another and modifying principle confronts that of continuity; we will call it the *principle of sufficient determination, or sufficient differentiation*.

The joint action of the two principles may be very well illustrated by a further analysis of the example cited. In order to deal with the phenomena exhibited in the change of color of light, the idea of the law of refraction must be retained, but with every particular color a particular index of refraction must be associated. We soon perceive that with every particular temperature also, a particular index of refraction must be associated; and so on.

In the end, this process leads to temporary contentment and satisfaction, the two things A and B being conceived as so connected that to every change of the one that can be observed at any moment there corresponds an appropriate change of the other. It may happen that both A and B are conceived as complexes of components, and that to every particular component of A a particular component of B corresponds. This occurs, for example, when B is a spectrum, and *A* the corresponding sample of a compound to be tested, in which case to every component part of the spectrum one of the components of the matter volatilized before the spectroscope is correlated, independently of the others. Only through complete familiarity with this relation can the principle of sufficient determination be satisfied.

2.

Suppose, now, that we are considering a color-sensation *B*, not in its dependence on *A*, the heated matter tested, but in its dependence on the elements of the retinal process, *N*. By doing this we change, not the kind, but only the direction of our point of view. None of the preceding observations lose their force, and the principles to be followed remain the same. And this holds good, of course, of all sensations.

Now, sensation may be analysed in itself, immediately, that is, psychologically (which was the course adopted by Johannes Müller), or the physical (physiological) processes correlated with it may be investigated according to the methods of physics (the course usually preferred by the modern school of physiologists), or, finally, the connexion of psychologically observable data with the corresponding physical (physiological) processes may be followed up a mode of procedure which will carry us farthest, since in this method observation is directed to all sides, and one investigation serves to support the other. We shall endeavor to attain this last-named end wherever it appears practicable.

This being our object, then, it is evident that the principle of continuity and that of sufficient determination can be satisfied only on the condition that with the same *B* (this or that sensation) we always associate the same *N* (the same nerve-process) and discover for every observable change of *B* a corresponding change of *N*. If *B* is psychologically analysable into a number of independent components, then we shall rest satisfied only on the discovery, in *N*, of equivalent components corresponding to these. If, on the other hand, properties or aspects have to be noticed in *B* which cannot appear in isolation, as, for instance, pitch and intensity in tones, we shall have to expect the same state of things in *N*. In a word, for all psychically observable details of *B* we have to seek the correlated physical details of *N*.

I do not of course maintain that a (psychologically) simple sensation cannot also be conditioned by very complicated circumstances. For the circumstances would hang together like the links of a chain and would not issue in a sensation, unless the chain extended to the nerve. But since the sensation may also appear in the form of a hallucination, namely when no physically conditioned circumstances are present outside the body, we see that a certain nervous process, as the final link in the chain, is the essential and immediate condition of the sensation. Now we cannot think of this immediate condition as being varied without conceiving of the sensation as being varied, and *vice versa*. For the connexion between this final link and the sensation we will regard the principle which we have laid down as valid.

3.

We may thus establish a guiding principle for the investigation of the sensations. This may be termed the *principle of the complete parallelism of the psychical and physical*. According to our fundamental conception, which recognizes no gulf between the two provinces (the psychical and the physical), this principle is almost a matter of course; but we may also enunciate it, as I did years ago, without the help of this fundamental conception, as a heuristic principle of research. [1]

The principle of which I am here making use goes further than the widespread general belief that a physical entity corresponds to every psychical entity and vice versa; it is much more specialized. The general belief in question has been proved to be correct in many cases, and may be held to be probably correct in all cases; it constitutes moreover the necessary presupposition of all exact research. At the same time the view here advocated is different from Fechner's conception of the physical and psychical as two different aspects of one and the same reality. In the first place, our view has no metaphysical background, but corresponds only to the generalized expression of experiences. Again, we refuse to distinguish two different aspects of an unknown tertium quid; the elements given in experience, whose connexion we are investigating, are always the same, and are of only one nature, though they appear, according to the nature of the connexion, at one moment as physical and at another as psychical elements. [2] I have been asked whether the parallelism between psychical and physical is not meaningless and a mere tautology, if the psychical and physical are not regarded as essentially different. The question arises from a misunderstanding of the analysis which I have given above. When I see a green leaf (an event which is conditioned by certain brain-processes) the leaf is of course different in its form and color from the forms, colors, etc., which I discover in investigating a brain, although all forms, colors, etc., are of like nature in themselves, being in themselves neither psychical nor physical. The leaf which I see, considered as dependent on the brain-process, is something psychical, while this brain-process itself represents, in the connexion of *its* elements, something physical. And the principle of parallelism holds good for the dependence of the former immediately given group of elements on the latter group, which is only ascertained by means of a physical investigation which may be extremely complicated (cp. p. 44).

4.

I have perhaps stated the principle in rather too abstract a form. A few concrete examples may now help to explain it. Wherever I have a sensation of space, whether through the sensation of sight or through that of touch, or in any other way, I am obliged to assume the presence of a nerve-process of the same kind in all cases. For all time-sensations, also, I must suppose like nerve-processes.

If I see figures which are the same in size and shape but differently colored, I seek, in connexion with the different color-sensations, certain identical space-sensations and corresponding identical nerve-processes. If two figures are similar (that is, if they yield partly identical space-sensations) then the corresponding nerve-processes also contain partly identical components. If two different melodies have the same rhythm, then, side by side with the different tone-sensations there exists in both cases an identical time-sensation with identical corresponding nerve-processes. If two melodies of different pitch are identical, then the tone-sensations as well as their physiological conditions, have, in spite of the different pitch, identical constituents. If the seemingly limitless multiplicity of color-sensations is susceptible of being reduced, by psychological analysis (self-observation), to six elements (fundamental sensations), a like simplification may be expected for the system of nerve-processes. If our system of space-sensations appears in the character of a threefold manifold, the system of the correlated nerve-processes will likewise present itself as such.

5.

This principle has, moreover, always been more or less consciously, more or less consistently, followed.

For example, when Helmholtz [3] assumes for every tone-sensation a special nerve-fibre (with its appurtenant nerve-process), when he resolves clangs, or compound sounds, into tone sensations, when he reduces the affinity of compound tones to the presence of like tone-sensations (and nerve-processes), we have in this method of procedure a practical illustration of our principle. It is only the application that is not complete, as will be later shown. Brewster, [4] guided by a psychological but defective analysis of color-sensations, and by imperfect physical experiments, [5] was led to the view that, corresponding to the three sensations, red, yellow, and blue, there existed likewise physically only three kinds of light, and that, therefore, Newton's assumption of an unlimited number of kinds of light, with a continuous series of refractive indices, was erroneous. Brewster might easily fall into the error of regarding green as a compound sensation. But had he reflected that color-sensations may occur entirely without physical light, he would have confined his conclusions to the nerve-process and left untouched Newton's assumptions in the province of physics, which are as well founded as his own. Thomas Young corrected this error, at least in principle. He perceived that an unlimited number of kinds of physical light with a continuous series of refractive indices (and wavelengths) was compatible with a small number of color-sensations and nerve-processes, - that a discrete number of color-sensations did answer to the continuum of deflexions in the prism (to the continuum of the space-sensations). But even Young did not apply the principle with full consciousness or strict consistency, wholly apart from the fact that he allowed himself to be misled, in his psychological analysis, by physical prejudices. Even he first assumed, as fundamental sensations, red, yellow, and blue, for which he later substituted red, green, and violet - misled, as Alfred Mayer, of Hoboken, has admirably shown, [6] by a physical error of Wollaston's. The direction in which the theory of color-sensation, which has reached a high degree of perfection through Hering, has still to be modified, was pointed out by me many years ago in another place.

6.

Here I will merely state shortly what I have to say concerning the treatment of the theory of color-sensation. We frequently meet with the assertion, in recent works, that the six fundamental color-sensations, white, black, red, green, yellow, blue, which Hering adopted, were first proposed by Leonardo da Vinci, and later by Mach and Aubert. From the very first it seemed to me highly probable, in view of the conceptions prevalent at this time, that the assertion was founded upon an error, as far as Leonardo da Vinci was concerned. Let us hear what he himself says in his *Book of Painting* (Nos. 254 and 255 in the translation of Heinrich Ludwig, *Quellenschriften zur Kunstgeschichte,* Vienna, Braumiiller, 1882, Vol. XVIII.). "254. Of simple colors there are six. The first of these is white, although philosophers admit neither white nor black into the number of colors, since the one is the cause of color, the other of its absence. But, *inasmuch as the painter cannot do without them,* we shall include these two also among the other colors and say that white in this classification is the first among the simple colors, yellow the second, green the third, blue the fourth, red the fifth, black the sixth. And the white we will let represent the light, without which one can see no color, the yellow the earth, the green the water, blue the air, red fire, and black the darkness which is above the element of fire, because in that place there is no matter or solid substance upon which the sunbeams can exert their force, and which as a result they might illumine." "255. Blue and green are not simple colors by themselves. For blue is composed of light and darkness, as, the blue of the air, which is made up of the most perfect black and perfectly pure white." "Green is composed of a simple and a composite color, namely, of yellow and blue." This will suffice to show that Leonardo da Vinci is concerned partly with observations concerning pigments, partly with conceptions of natural philosophy, but not with the subject of fundamental color-sensations. The many remarkable and subtle scientific observations of all sorts which are contained in Leonardo da Vinci's book lead to the conviction that the artists, and among them especially he himself, were the true forerunners of the great scientists who came soon afterwards, These men were obliged to understand nature in order to reproduce it agreeably; they observed themselves and others in the interest of pure pleasure. Yet Leonardo was far from being the author of all the discoveries and inventions which Groth, for example (*Leo-*

nardo da Vinci ah Ingenieur und Philosoph, Berlin, 1874), ascribes to him. [7] My own scattered remarks concerning the theory of color-sensations were perfectly clear. I assumed the fundamental sensations white, black, red, yellow, green, blue, and six different corrresponding (chemical) processes (not nerve-fibres) in the retina. (Compare *Reichert's und Dubois' Archiv,* 1865, p. 633, et seq.) As a physicist, I was of course familiar with the relation of the complementary colors. My conception, however, was that the two complementary processes together excited a new the white process. (*Loc. cit.,* p. 634.) I gladly acknowledge the great advantages of Hering's theory. They consist for me in the following. First, the black process is regarded as a reaction against the white process; I can appreciate all the better the facilitation involved in this, as it was just the relation of black and white that for me presented the greatest difficulty. Further, red and green, as also yellow and blue, are regarded as antagonistic processes which do not produce a new process, but mutually annihilate each other. According to this conception white is not subsequently produced but is already present beforehand, and still survives on the annihilation of a color by the complementary color. The only point that still dissatisfies me in Hering's theory is that it is difficult to perceive why the two opposed processes of black and white may be simultaneously produced and simultaneously felt, while such is not the case with red-green and blue-yellow. This objection has been partly removed by a further development of Hering's theory. [8] The full explanation of this relation lies undoubtedly in the proof, which W. Pauli has provided, that certain processes in colloidal and in living substances can be reversed by opposite processes along the same path, or "homodromously," as in *A*, while other processes can only be reversed by opposite processes along a different path, or "heterodromously," as in *B*. [9] I myself showed long ago that certain sensations are related to one another as positive and negative magnitudes (*e.g.*, red and green), while others do not stand in this relation (*e.g.*, white and black). [10] Now all difficulties are reconciled if we suppose with Pauli that the opposed processes as assumed by Hering, which correspond to the first pair, are homodromous, and that the processes underlying the second pair are heterodromous. [11]

7.

The examples adduced will suffice to explain the significance of the above-enunciated principle of inquiry, and at the same time to show that this principle is not entirely new. In formulating the principle, years ago, I had no other object than that of making quite clear to my own mind a truth which I had long instinctively felt.

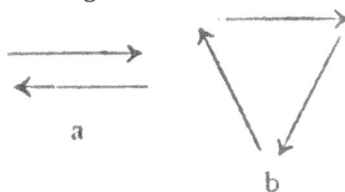

It seemed to me a simple and natural, nay, an almost self-evident supposition, that similarity must be founded on a partial likeness or identity, and that consequently, where sensations were similar, we had to look for their common identical constituents and for the corresponding common physiological processes. I wish, however, to make it quite clear to the reader that this view by no means meets with universal agreement. We constantly find it maintained in philosophical books that similarity may be observed without there being any question at all of such identical constituents. Thus a physiologist [12] can speak as follows of the principle under discussion: "The application of this principle to the above problems leads him (Mach) to ask. what is the physiological factor that corresponds to the qualities thus postulated? Now it seems to me that, of all axioms and principles, none is more doubtful, none is exposed to greater misunderstandings than is this principle. If it is nothing more than a periphrasis for the so-called principle of parallelism, then it cannot be considered either new or particularly fruitful, and it does not deserve the importance that is attached to it. If, on the other hand, it is intended to mean that a definite element or constituent of a physiological event must correspond to everything which we can distinguish as having some sort of psychological unity, to every relation, to every form, in a word to everything that we can denote by a general conception, then this formulation can only be characterized as dubious and misleading." And I am taken as holding that the principle in question must be understood in this last "dubious and misleading" sense. I must leave it entirely to the reader to choose whether he will accompany me any further and enter with me on that preliminary stage of inquiry which is clearly denned by means of our principle, or whether, bowing to the authority of my opponents, he will turn back and satisfy himself merely with considering the difficulties which confront him. If he chooses the former alternative, he will, I hope, discover, that when simpler cases have been disposed of, the difficulties in cases of deeper-lying abstract similarity no longer appear in such a formidable light as before. All I will add at present is, that in these more] complicated cases of similarity the similarity arises not from the presence of one common element, but from a common system of elements, as I shall explain at length in connexion with conceptual thinking (Cf. Chap. XIV.).

8.

As we recognize no real gulf between the physical and the psychical, it is a matter of course that, in the study of the sense-organs, general physical as well as special biological observations may be employed. Much that appears to us difficult of comprehension when we draw a parallel between a sense-organ and a physical apparatus, is rendered quite obvious in the light of the theory of evolution, simply by assuming that we are concerned with a living organism with particular memories, particular habits and manners, which owe their origin to a long and eventful race-history. The sense-organs themselves are a fragment of soul; they themselves do part of the psychical work, and hand over the completed result to consciousness. I will here briefly put together what I have to say on this subject.

9.

The idea of applying the theory of evolution to physiology in general, and to the physiology of the senses in particular, was advanced, prior to Darwin, by Spencer (1855). It received an immense impetus through Darwin's book *The Expression of the Emotions*. Later, P. R. Schuster (1879) discussed the question whether there were "inherited ideas" in the Darwinian sense. I, too, expressed myself in favor of the application of the idea of evolution to the theory of the sense-organs (*Sitzungsberichte der Wiener Akademie,* October 1866). One of the finest and most instructive discussions, in the way of a psychologico-physiological application of the theory of evolution, is to be found in the Academic Anniversary Address of Hering, *On Memory as a General Function of Organized Matter;* 1870 (English translation, Open Court Publishing Company, Chicago, 1913). As a fact, memory and heredity almost coincide in one concept if we reflect that organisms, which were part of the parent-body, emigrate and become the basis of new individuals. Heredity is rendered almost as intelligible to us by this thought as, for example, is the fact that Americans speak English, or that their state-institutions resemble the English in many respects, etc. The problem involved in the fact that organisms possess memory, a property which is apparently lacking to inorganic matter, is, of course, not affected by these considerations, but still exists (cp. Chapters V., XL). If we want to avoid criticizing Hering's theory unfairly, we must observe that he uses the conception of memory in a rather broad sense. He perceived the affinity between the lasting traces imprinted on organisms by their racial history, and the more evanescent impressions which the individual life leaves behind it in consciousness. He recognizes that the spontaneous reappearance, in response to a slight stimulus, of a process which has once been set up, is essentially the same event, whether it can be observed within the narrow framework of consciousness or not. The perception of this common feature in a long series of phenomena is an essential step in advance, even though this fundamental feature itself still remains unexplained. [13] Recently Weismann (*Ueber die Dauer des Lebens,* 1882) has conceived death as a phenomenon of heredity. This admirable book, also, has a very stimulating effect. The difficulty which might be found in the fact that a characteristic should be inherited which can make its appearance in the parent-organism only after the process of inheritance is ended, lies probably only in the manner of statement. It disappears when we consider that the power of the somatic cells to multiply can increase, as Weismann shows, at the cost of the increase of the germ-cells. Accordingly, we may say that greater length of life on the part of the cell-society and lessened propagation are two phenomena of adaptation which mutually condition each other. While a Gymnasium student, I heard it stated that plants from the Southern Hemisphere bloom in our latitudes, when it is spring in their native place. I recall clearly the mental shock which this communication caused me. If it is true, we may actually say that plants have a sort of memory, even though it be admitted that the chief point involved is the periodicity of the phenomena of life. The so-called reflex movements of animals may be explained in a natural manner as phenomena of memory outside the organ of consciousness. I was a witness of a very remarkable phenomenon of this kind - in 1865, I think - with Rollett, who was experimenting with pigeons whose brains had been removed. These birds drink whenever their feet are placed in a cold liquid, whether the liquid is water, mercury, or sulphuric acid. Now since a bird must ordinarily wet its feet when it seeks to quench its thirst, the view arises quite naturally that we have here a habit adapted to an end, which is conditioned by the mode of life and fixed by inheritance, and which, even when consciousness is eliminated, takes place with the precision of clockwork on the application of the stimulus appropriate to its excitation. Goltz, in his wonderful book *Die Nervencentren des Frosches,* 1869, and in later writings, has described many phenomena of the sort. - I will take this opportunity of mentioning some further observations which I recall with a great deal of pleasure. In the autumn vacation of 1873, my little boy brought me a sparrow a few days old, which had fallen from its nest, and wanted to bring it up. But the matter was not so easy. The little creature could not be induced to swallow, and would certainly soon have succumbed to the indignities that would have been unavoidable in feeding it by force. I then fell into the following train of thought: "Whether or not the Darwinian theory is correct, the new-born child would certainly perish if it had not the specially formed organs and inherited impulse to suck, which are brought into activity quite automatically and mechanically by the appropriate stimulus. Something similar (in another form) must exist likewise in the case of the bird." I exerted myself to discover the appropriate stimulus. A small insect was stuck upon a sharp stick and swung rapidly about the head of the bird. Immediately the bird opened its bill, beat its wings, and eagerly devoured the proffered food. I had thus discovered the right stimulus for setting the impulse and the automatic movement free. The creature grew perceptibly stronger and greedier, it began to snatch at the food, and once seized an insect that had accidentally fallen from the stick to the table; from that time on it ate, without ceremony, of itself. In proportion as its intellect and memory developed, a smaller portion of the stimulus was required. On reaching .independence, the creature took on, little by little, all the characteristic ways of sparrows, which it certainly had not learnt by itself. By day, with its intellect awake, it was very trustful and friendly. In the evening, other phenomena were regularly exhibited. It grew timid. It always sought out the highest places in the room, and would become quiet only when it was prevented by the ceiling from going higher. Here again we have an inherited habit adapted to an end. On the coming of darkness, its demeanour changed totally. When approached, it ruffled its feathers, began to hiss, and showed every appearance of terror and real physical fear of ghosts. Nor is this fear without its reasons and its purpose in a creature which, under normal circumstances, may at any moment be devoured by some monster.

This last observation strengthened me in an opinion already formed, that my children's terror of ghosts did not have its source in nursery tales, which were carefully excluded from them, but was innate. One of my children would regard with anxiety an arm-chair, which stood in the shadow; another carefully avoided, in the evening, a coalscuttle by the stove, especially when this stood with the lid open, looking like gaping jaws. The fear of ghosts is the true mother of religions. Neither scientific analysis nor the careful historical criticism of a David Strauss, as applied to myths, which, for the strong intellect, are refuted even before they are invented, will all at once do away with and banish these things. A motive which has so long answered, and in a measure still answers, to actual economic needs (fear of something worse, hope of something better), will long continue to exist in mysterious and uncontrollable instinctive trains of thought. Just as the birds on uninhabited islands (according to Darwin) learn the fear of man only after the lapse of generations, so we shall unlearn, only after many generations, that useless habit known as the creeping of flesh. Every presentation of Faust may teach us the extent to which we are still in secret sympathy with the conceptions of the age of witchcraft. The exact knowledge of nature and of the conditions of this life gradually becomes more useful to man than fear of the unknown. And in time the most important thing of all for him is to be on his guard against his fellow-men who want to oppress him violently or abuse him treacherously by misleading his understanding and emotions. I will here relate one other curious observation, for the knowledge of which I am indebted to my father (an enthusiastic Darwinian and in the latter part of his life a landed proprietor in Carniola). My father occupied himself much with silk-culture, raised the yama-mai in the open oak-woods, etc. The ordinary mulberry silkworm has, for many generations, been raised indoors, and has consequently become exceedingly helpless and dependent. When the time for passing into the chrysalitic state arrives, it is the custom to give the creatures bundles of straw, upon which they spin their cocoons. Now it one day occurred to my father not to prepare the usual bundles of straw for a colony of silk-worms. The result was that the majority of the worms perished, and only a small portion, the geniuses (those with the greatest power of adaptation) spun their cocoons. Whether, as my sister believes she has observed, the experiences of one generation are utilized, in noticeable degree, in the very next generation, is a question which probably requires to be left to further investigation. The experiments made by C. Lloyd Morgan (*Comparative Psychology*, London, 1894) with young chickens, ducks, etc., show that, at any rate in the case of the higher animals, scarcely anything is innate but the reflexes. The newly-hatched chick at once begins to peck with great assurance at everything that it sees; but it has to learn what is suitable to pick up by its individual experience. The simpler the organism, the smaller the part played by individual memory. From all these remarkable phenomena we need derive no mysticism of the Unconscious. A memory reaching beyond the individual (in the broader sense defined above) renders them intelligible. A psychology in the Spencer-Darwinian sense, founded upon the theory of evolution, but supported by detailed positive investigation, would yield richer results than all previous speculation has done. These observations and reflections had long been made and written down when Schneider's valuable work, *Der thierische Wille,* Leipzig, 1880, which contains many that are similar, made its appearance. I agree with the details of Schneider's discussions (in so far as they have not been made problematical by Lloyd Morgan's experiments) almost throughout, although his fundamental conceptions in the realms of natural science with regard to the relation of sensation and physical process, the significance of the survival of species, etc., are essentially different from mine, and although I hold, for instance, the distinction between sensation-impulses and perception-impulses to be quite superfluous. An important revolution in our views on heredity may perhaps be produced by Weismann's work, *Ueber die Vererbung,* Jena, 1883 (English translation, *Essays on Heredity and Kindred Biological Problems,* Oxford, The Clarendon Press, 1889). Weismann regards the inheritance of traits acquired by use as highly improbable, and finds in chance variation of the germ-elements and in the selection of the germ-elements the most important factors. Whatever attitude we adopt towards Weismann's theories, the discussion initiated by him must contribute to the elucidation of these questions. No one will refuse to recognize the almost mathematical acuteness and depth of the way in which he states the problem, and it cannot be denied that his arguments have much force. He makes, for instance, the extremely suggestive remark that it is impossible that the peculiar and unusual forms of sexless ants, which must apparently be referred to use and adaptation, and which moreover deviate so remarkably from the forms of ants that are capable of propagation, should be produced by inheritance of characteristics acquired by use. [14] That the germ-elements themselves may be altered by external influences appears to be clearly shown by the formation of new races, which maintain themselves as such, transmit their racial traits by inheritance, and are themselves, again, capable of transformation, under other circumstances. Accordingly, some influence must certainly be exerted on the germ-plasm by the body which envelops it (as Weismann himself admits). Thus an influence of the individual life upon its descendents can certainly not be entirely excluded, even although a direct transmission to the descendents of the results of use in the individual can (according to Weismann) no longer be expected. In entertaining the notion that the germ-elements vary accidentally, we must bear in mind that chance is not a principle of action. When periodic circumstances of different kinds and different periodicities coincide in accordance with definite causal laws, the circumstances overlap in such a way that in any particular case it is impossible to see that any law is involved. But the law reveals itself with the lapse of a long enough time, and permits us to calculate on certain average values or probabilities of effects. [15] Without some such principle of action, chance or probability is meaningless. And what principle of action can be conceived as exercising more influence on the variation of the germ-elements than the body of the parent? Personally I cannot understand how it is possible that the species should succumb to the influence of varying circumstances, and yet that these circumstances should not affect the individual. Moreover, I am certain that I myself vary with every thought, every memory, every experience; all these factors undoubtedly change my whole physical behaviour. [16]

Although it is scarcely necessary, I should like to add explicitly that I regard the theory of evolution, in whatever form, as a working scientific hypothesis, capable of being modified and of being made more precise, which is valuable in so far as it facilitates the provisional understanding of what is given in experience. I have been a witness of the powerful impetus which Darwin's work gave in my time not merely to biology, but to all scientific enquiry, and it is not likely that I should underestimate the value of the theory of evolution. But I would not quarrel with anyone who should rate its value very low. As long ago as 1883 and 1886 I dwelt on the necessity of advancing by means of more precise conceptions obtained by the study of biological facts for their own sakes. [17] Thus I am by no means committed to a refusal to understand investigations such as those of Driesch. But whether Driesch's criticism of my attitude towards the theory of evolution is justified, [18] I leave to anyone to decide who, even after this criticism, still cares to be at the pains of reading my works.

10.

Teleological conceptions, as aids to investigation, are not to be shunned. It is true, our comprehension of the facts of reality is not enhanced by referring them to an unknown World-Purpose, itself problematical, or to the equally problematical purpose of a living being. Nevertheless, the question as to the value that a given function has for the existence of an organism, or as to what are its actual contributions to the preservation of the organism, may be of great assistance in the comprehension of this function itself. [19] Of course we must not suppose, on this account, as many Darwinians have done, that we have "mechanically explained" a function, when we discover that it is necessary for the survival of the species. Darwin himself is doubtless quite free from this short-sighted conception. By what physical means a function is developed, still remains a physical problem; while the how and why of an organism's voluntary adaptation continues to be a psychological problem. The preservation of the species is only one, though an actual and very valuable, point of departure for inquiry, but it is by no means the last and the highest. Species have certainly been destroyed, and new ones have as certainly arisen. The pleasure-seeking and pain-avoiding will, [20] therefore, is directed perforce beyond the preservation of the species. It preserves the species when it is advantageous to do so, and destroys it when its survival is no longer advantageous. Were it directed merely to the preservation of the species, it would move aimlessly about in a vicious circle, deceiving both itself and all individuals. This would be the biological counterpart of the notorious "perpetual motion" of physics. The same absurdity is committed by the statesman who regards the state as an end in itself.

[1] Compare my paper, *Ueber die Wirkung der räumlichen Vertheilung des Lichtreizes aufdie Netzhaut* (*Sitzungsberichte der Wiener Akadamie*, Vol. LII., 1865); further *Reichert's und Dubois' Archiv*, 1865, p. 634, and *Grundlinien der Lehre von den Bewegungsempfindungen* (Leipzig, Engelmann, 1875, p. 63). The principle is also implicitly contained in an article of mine in Fichte's *Zeitschrift für Philosophie* (Vol. XLVI., 1865, p. 5), which is printed also in my *Popular Scientific Lectures,* Chicago, Open Court Publishing Co.

[2] For the various aspects of the problem of parallelism, see C. Stumpf's address to the Psychological Congress at Munich (Munich, 1897); G. Heymans, "Zur Parallelismusfrage," *Zeitschrift für Psychologie der Sinnesorgane,* Vol. XVII.; O. Külpe, *Ueber die Beziehung zwischen körperlichen und seelischen Vorgängen, Zeitschrift für Hypnotismus,* Vol. VII., J. von Kries, *Über die materiellen Grundlagen der Bewusstseinserscheinungen,* Freiburg im Breisgau, 1898; C. Hauptmann, *Die Metaphysik in der Psychologie,* Dresden, 1893.

[3] Helmholtz, *Die Lehre von den Tonempfindungen,* Brunswick, Vieweg, 1863. English translation by Alex. J. Ellis, London, Longmans, Green, & Co., 2nd edition, 1885.

[4] Brewster, *A Treatise on Optics,* London, 1831. Brewster regarded the red, yellow, and blue light as extending over the whole solar spectrum, though distributed there with varying intensity, so that, to the eye, red appears at both ends (the red and the violet), yellow in the middle, and blue at the end of greater refrangibility.

[5] Brewster believed that he was able to alter by absorption the nuances of the spectrum colors regarded by Newton as simple a result which, if correct, would really destroy the Newtonian conception. He experimented, however, as Helmholtz (*Physiological Optics*) has shown, with an impure spectrum.

[6] *Philosophical Magazine,* February 1876, p. 111. Wollaston was the first to notice (1802) the dark lines of the spectrum, later named after Fraunhofer, and believed that he saw his narrow spectrum divided by the strongest of these lines into a red, a green, and a violet part. He regarded these lines as the dividing lines of the physical colors. Young took up this conception, and substituted for his fundamental sensations red, yellow, and blue, the colors red, green, and violet. Thus, in his first conception, Young regarded green as a composite sensation, in his second, both green and violet as simple. The questionable results which psychological analysis may thus yield, are well calculated to destroy belief in its usefulness in general. But we must not forget that there is no principle in the application of which error is excluded. Here, too, practice must determine. The circumstance that the physical conditions of sensation almost always give rise to composite sensations, and that the components of sensation seldom make their appearance separately, renders psychological analysis very difficult. Thus, green is a simple sensation; a given pigment or spectrum-green, however, will as a rule

excite also a concomitant yellow or blue sensation, and thus favor the erroneous idea (based upon the results of pigment-mixing) that the sensation of green is compounded of yellow and blue. Careful physical study, therefore, is also an indispensable requisite of psychological analysis. On the other hand, physical observation must not be overestimated. The mere observation that a yellow and blue pigment mixed, yield a green pigment, cannot by itself determine us to see yellow and blue in green, unless one or the other color is actually contained in it. Certainly no one sees yellow and blue in white, although, as a fact, spectrum-yellow and spectrum-blue mixed give white.

[7] Marie Herzfeld, *Leonardo da Vinci, Auswahl aus den veröffentlichten Handschriften,* Leipzig, 1904.

[8] *Zur Lehre vom Lichtsinne,* Vienna, 1878, p. 122. Cp. also my paper, previously cited, in the *Sitzungsberichte der Wiener Akademie,* Vol. LII., 1865, October.

[9] W. Pauli, *Der Kolliodale Zustand und die Vorgänge in der lebendigen Substanz,* Brunswick, Vieweg, 1902, pp. 22, 30.

[10] *Grundlinien der Lehre von den Bewegungsempfindungen,* 1895, PP57 , sqq.

[11] A recent exposition of Hering's views will be found in Graefe-Saemisch's *Handbuch der ges. Augenheilkunde,* Leipzig, 1905, Vol. III.

[12] J. Von Kries, *Ueber die materiellen Grundlagen der Bewusstseinserscheinungen,* Freiburg im Breisgau, 1898.

[13] R. Semon, *Die Mneme,* Leipzig, 1904.

[14] But perhaps the powerful mandibles of the sexless ants are the *original* acquisition of the species, and merely appear in an atrophied form in the individuals to whom propagation of the race is confined.

[15] *Vorlesungen über Psychophysik, Zeitschrift für prakt. Heilkunde,* pp. 148, 168, 169, Vienna, 1863.

[16] *Popular Scientific Lectures,* Chicago, Open Court Publishing Co.

[17] Cp. *Popular Scientific Lectures,* and *Analyse der Empfindungen,* 1886, pp. 34, sqq.

[18] Driesch, *Die organisatorischen Regulationen,* pp. 165, sqq., 1901.

[19] Such teleological conceptions have often been useful and instructive to me. The remark, for example, that a visible object under varying intensity of illumination can be recognized as the same only when the sensation excited depends on the ratio of the illumination-intensities of object and surroundings, makes intelligible a whole train of organic properties of the eye. (Cp. Hering in Graefe-Saemisch's *Handbuch der Augenheilkunde,* Vol. III., Ch. 12, pp. 13, sqq.) In this way we understand also, how the organism, in the interest of its survival, was obliged to adjust itself to the requirement mentioned and to adapt itself to feel the ratios of light-intensity. The so-called law of Weber, or the fundamental psycho-physical formula of Fechner, thus appears not as something fundamental, but as the explicable result of organic adjustments. The belief in the universal validity of this law is, naturally, herewith relinquished. I have given the arguments on this point in various papers. (*Sitzungsberichte der Wiener Akademie,* Vol. LII., 1865; *Vierteljahrsschrift für Psychiatrie,* Neuwied and Leipzig, 1868; *Sitzungsberichte der Wiener Akademie,* Vol. LVII., 1868). In the last-named paper, proceeding from the postulate of the parallelism between the psychical and the physical, or, as I then expressed myself, from the proportionality between stimu-

lus and sensation, I abandoned the metrical formula of Fechner (the logarithmic law), and brought forward another conception of the fundamental formula, the validity of which for light-sensation I never disputed. This is apparent beyond all doubt from the way in which that paper is worked out. Thus one cannot say, as Hering has done, that I everywhere take the psychophysical law as my foundation, if by this is understood the *metrical formula*. How could I have maintained the *proportionality* between stimulus and sensation at the same time with the *logarithmic* dependence? It was sufficient for me to render my meaning clear; to criticize and contest Fechner's law in detail, I had, for many obvious reasons, no need. Strictly speaking I consider the expression "proportionality" also to be inappropriate, since there can be no question of an actual measurement of the sensations; all that can be done is to characterize them exactly and make an inventory of them by numerical means. Cp. what I have said about the characterization of states of heat (*Prinzipien der Wärmelehre,* p. 56).

[20] Schopenhauer's conception of the relation between Will and Force can quite well be adopted without seeing anything metaphysical in either.

V. Physics and Biology - Causality and Teleology

1.

IT often happens that the development of two different fields of science goes on side by side for long periods, without either of them exercising an influence on the other. On occasion, again, they may come into closer contact, when it is noticed that unexpected light is thrown on the doctrines of the one by the doctrines of the other. In that case a natural tendency may even be manifested to allow the first field to be completely absorbed in the second. [1] But the period of buoyant hope, the period of over-estimation of this relation which is supposed to explain everything, is quickly followed by a period of disillusionment, when the two fields in question are once more separated, and each pursues its own aims, putting its own special questions and applying its own peculiar methods. But on both of them the temporary contact leaves abiding traces behind. Apart from the positive addition to knowledge, which is not to be despised, the temporary relation between them brings about a transformation of our conceptions, clarifying them and permitting of their application over a wider field than that for which they were originally formed.

2.

We are living at present in such a period of complicated cross-relations, and the consequent fermentation of ideas gives rise to very remarkable phenomena. While many physicists are concerned to purify physical conceptions by psychological, logical and mathematical methods, other physicists, mistrustful of this tendency, and more philosophical than the philosophers themselves, are coming forward as advocates of the old metaphysical conceptions which the philosophers have already largely abandoned. Philosophers, psychologists, biologists, and chemists, all make the most widely extended applications of the principle of energy and of other physical conceptions, with a freedom which the physicist would hardly venture to use in his own field. We may almost say that the customary roles of the special departments have been interchanged. The success of this movement may be partly positive and partly negative, but in any case the result of it will be a more precise determination of our conceptions, a more accurate delimitation of the sphere to which they apply, and a clearer idea of the difference and the affinity between the methods of the departments in question.

3.

We are here concerned in particular with the relations between the physical and biological fields in the broadest sense. The distinction between effective causes and final causes, or ends, dates from Aristotle. It has been generally assumed that physical phenomena are throughout determined by effective causes, and biological phenomena by final causes. The acceleration of a body, for example, is determined solely by the effective causes, by the circumstances of the movement, such as the presence of other gravitating magnetic or electrical bodies. Up to the present we are not able to deduce the development of a growing animal, or the development of a plant in its peculiar determinate forms, or the instinctive actions of an animal, from effective causes alone; but such facts as these can be at any rate partially understood when we take into consideration the purpose of self-preservation under the particular circumstances of the organism's life. Whatever theoretical reservations we may make as to the application to biology of the conception of purpose, it would certainly be perverse, in a field where the "causality" theory still affords such imperfect explanations, to refuse to make use of the clues which a consideration of purpose puts into our hands. I do not know what it is that compels the caterpillar of the hawk-moth to spin a cocoon with a bristly flap opening outwards, but I see that such a cocoon exactly corresponds to the purpose of preserving the caterpillar's existence. I am far from being able to understand "causally" the numerous remarkable phenomena, described and studied by Reimarus and Autenrieth, of the development and instinctive action of animals; but I understand them in the light of the purpose of preserving the animal's existence in the particular conditions of life involved. These phenomena consequently merit attention; they become fused in the picture which we form of the life of the organism as ineffaceable constituents; and it is only through them that that picture can be rounded out into a united and connected whole. But it is only quite recently, and particularly through the investigations of Sachs in the physiology of plants and, in animal physiology, through the work of Loeb on geotropism, heliotropism, stereotropism, etc., that the relations between growth and instinct have been really explained in such a way that we can begin to conceive these relations also as "causal." History testifies in a manner that cannot be gainsaid to the utility of the conception of purpose in biological research. Consider only Kepler's investigation of the eye. It was impossible for him, in view of the purpose of the eye, namely clear vision at different distances, to doubt the existence of accommodation; but it was not until 150 years later that the processes which effect the accommodation were really discovered. Harvey discovered the circulation of the blood in the course of an attempt to make clear to himself the problematical purpose of the position of the valves of the heart and veins.

Even when a department of facts has been completely explained teleologically, the need to understand it "causally" still persists. The belief in the completely different nature of the two departments we are considering, in virtue of which one is to be understood only in a causal sense and the other only in a teleological sense, is not justified. A complex of physical facts is something simple, or at any rate in many cases can, at will, be experimentally represented in such a simple form that the immediate relations between its parts become visible. Now supposing that we have done sufficient work in this department to have enabled us to acquire, as regards the nature of these relations, conceptions which we think correspond to the facts universally, then we are logically bound to expect that any particular fact which may present itself will correspond to these conceptions. But this implies no necessity in nature. [2] It is in this that "causal" understanding consists. A biological factual complex, on the other hand, is compounded in such a way that the immediate relations between its parts cannot be taken in at a glance. Accordingly we are satisfied if we are able to represent as being connected with one another prominent parts of the complex which are not immediately connected. But the intellect which has been trained to familiarity with the simpler causal relation finds, in the absence of the intermediate links, difficulties which it tries to remove as best it can, either by trying to discover these intermediate links, or by grasping at the hypothesis of a quite new kind of connecting relations. The latter alternative is unnecessary, if we regard our knowledge as imperfect and provisional, and reflect that in the department of physics absolutely analogous cases arise. The scientists of antiquity, indeed, did not draw this precise distinction between the two departments. Aristotle, for example, conceives heavy bodies as seeking out their position; Hero thinks that, from motives of economy, nature conducts light by the shortest paths and in the shortest times. These inquirers thus set up no such definite boundary between the physical and the biological. Moreover, by an imperceptible modification in our thought, we can formulate every teleological question in such a way as completely to exclude the conception of purpose. The eye sees clearly at different distances; the apparatus of dioptrical vision must therefore be capable of change; in what does this change consist? The valves of the heart and veins all open in the same direction; this being so, the circulation of the blood can only take place in one direction. Is this the fact? The modern theory of evolution has made this sober method of thought its own. Even in the very advanced parts of physics, on the other hand, we come across considerations having a great affinity with those of the biological sciences. The investigation, for example, of the possibility, under certain conditions, of stationary vibrations (*i.e.*, vibrations which can maintain themselves) has for some time past been in an advanced state. It is only quite recent work, however, which has made clear the manner in which they arise. [3] We explain the movement of light along the shortest paths by means of a selection of the affective paths. At present the conceptions used by chemists are even closer to those of the biologist. According to these conceptions all possible combinations are formed by the resolution of elements; but combinations which cannot be resolved, and have greater power of resistance to new attacks, get the better of the others and survive. It thus would appear that there is as yet no necessity to assume a fundamental difference between teleological and causal methods of investigation. The former is simply a provisional method.

5.

To confirm this conclusion in greater detail, let us return to the various conceptions of causality. The old traditional conception of causality is of something perfectly rigid: a dose of effect follows on a dose of cause. A sort of primitive, pharmaceutical conception of the universe is expressed in this view, as in the doctrine of the four elements. The very word "cause" makes this clear. The connections of nature are seldom so simple, that in any given case we can point to one cause and one effect. I therefore long ago proposed to replace the conception of cause by the mathematical conception of function, that is to say, by the conception of the dependence of phenomena on one another, or, more accurately, the dependence of the characteristics of phenomena on one another. [4] This conception is capable of any extension or limitation that may be desired, according to what is required by the facts under investigation.

Perhaps, therefore, great importance need not be attached to the objections that have been raised against it. [5] Consider, as a simple example, the relation of gravitating masses. If a mass B comes into opposition to a mass A, a movement of A towards B follows. This is the old formula. But, if we consider the matter more accurately, we see that the masses $A\ B\ C\ D$...determine mutual accelerations in one another, accelerations which, therefore, are given as soon as the masses are posited. The accelerations allow us to infer the velocities which will be attained at some future moment. Hence also the positions of $A\ B\ C\ D$...are determined for every moment. But the physical measurement of time is based in its turn upon a measurement of space, namely the rotation of the earth. We are thus ultimately left with a mutual dependence of positions on one another. Thus, even in this simplest case, the old formula is incapable of embracing the multiplicity of the relations that exist in nature. Similarly in other cases everything is resolved into relations of mutual dependence, [6] as to the form of which nothing of course can be said beforehand, since this is a question which can only be settled by specialized inquiry. With a relation of mutual dependence change is only possible when some group of the related elements can be regarded as an independent variable. Consequently, though it may be possible to complete in detail the picture of the world in a scientifically determined manner when a sufficient part of the world is given, yet science cannot tell us what the total result of the world process will be.

Given a well-defined mechanical system (defined, say, by central forces), with its positions and velocities, its configuration is determined as a function of time. We know its configuration for any time we like before and after the time of commencement, and can thus prophesy both backwards and forwards. This can only happen in both cases when no disturbance intervenes from outside, when, that is to say, the system can be regarded as in a certain sense a closed system. We cannot, indeed, regard any system as being completely isolated from the rest of the world, inasmuch as the determination of time, and consequently also that of the velocities, presupposes dependence on a parameter which is determined by the path traversed by some body, such as a planet, lying outside the system. The actual dependence, even though it be not an immediate dependence of all processes on the position of one body, guarantees to us the interconnection of the whole world. Analogous considerations hold for any physical system, even when it is not conceived as a mechanical system also. All accurately and clearly recognized relations may be regarded as mutual relations of simultaneity.

Let us consider, by way of contrast, the popular conception of cause and effect. Let S in Fig. 1, b, represent the sun, which illumines a body K placed in any medium whatever. Then the sun, or the heat of the sun, is the cause of a rise in temperature in the body K. The rise follows regularly on the illumination of K. On the other hand, the body K) or the change in its temperature, cannot be regarded as the cause of the change in the sun's temperature, as would actually be the case if S and K stood alone in an immediate relation to one another. The two changes would then be simultaneous, and would mutually determine one another. The reason that this is not so is to be sought in the intermediate links, the elements, A B, of the medium, which determine changes not only in K, but also in other elements, and in their turn are determined by these latter. Thus K stands in relations of mutual determination with innumerable elements, and only a vanishing portion of the light it reflects finds its way back to the sun. It is in analogous circumstances that we must look for the reason why a body K throws an image upon the retina N, and sets up a visual sensation E, from which a memory remains behind, although the memory does not restore either the retina N or the whole body K. The principal advantage for me of the notion of function over that of cause lies in the fact that the former forces us to greater accuracy of expression, and that it is free from the incompleteness, indefiniteness and one-sidedness of the. latter. The notion of cause is, in fact, a primitive and provisional way out of a difficulty. Every modern man of science must, I think, feel this, when, for instance, he glances at J. S. Mill's discussion of the methods of experimental inquiry. If he were to try to apply these methods he would never get beyond the most rudimentary results. The range of spatial and temporal functional relations within which our conjectures operate, may be one of which the limits lie very far apart: starting from the present we may try to prophesy into the distant future or past, and we may make fortunate guesses. But, the greater the distance, the less secure must the basis of our reasoning be. It is therefore, without prejudice to the greatness of Newton's conception of action at a distance, a very important step in advance that modern physics, wherever it can, requires that due consideration should be paid to spatial and temporal continuity.

6.

It might seem from what has been said that, both in physics and biology, the notion of function was all that we wanted, and that this notion would prove equal to all requirements. We need not be alarmed by the great difference in point of view displayed by the two sciences. Quite closely related groups of physical phenomena, as for example frictional electricity and galvanic electricity, look so different that at first sight it might seem impossible to expect that the two should be capable of reduction to the same fundamental facts. The magnetic and chemical phenomena, which are scarcely observable in the case of frictional electricity, and could only with difficulty have been discovered there at all, are extremely prominent in galvanic electricity; whereas, contrariwise, it is only in the case of the former that ponderomotive phenomena and phenomena of tension present themselves easily and unsought for. Now it is well known that each of these two studies supplements and throws great light on the other; so much so that we have reached the point of discovering the chemical nature of frictional electricity by means of galvanic electricity.

An analogous relation also holds between physics and biology. Both contain the same fundamental facts; but many sides of these facts come to light only in one of them, while many other sides are only noticeable in the other, so that physics, standing side by side with biology, can afford help to and throw light upon biology, and vice versa. No one can deny that the application of physics to biology has accomplished much; but against these achievements we have to set other cases in which it was reserved for biology to bring to light new physical facts (galvanism, Pfeffer's cells, etc.). Physics will accomplish much more in biology, if only she will submit to have additions made to her by the latter.

7.

If anyone familiar with the physical sciences alone were to turn to the study of biology, and thereupon were to suppose that an animal grows special organs which it finds ready to be applied to some useful purpose at a later stage of its life, or that it can perform instinctive actions which it cannot have learnt and which can only inure to the advantage of future members of the race, or that it adapts its coloration to its environment in order to avoid possible future enemies, on such suppositions he would in fact easily arrive at the assumption that quite peculiar factors were here at work. One excellent reason why this mysterious operation of the future at a distance cannot be compared with any physical relation, is that the operation does not take place exactly and without exception, since many organisms prepare themselves for a later stage of their life, but are destroyed before attaining it. It is impossible to regard something which is not determined for ourselves, or is only partially determined, namely the uncertain past or the uncertain future, as being the determining factor in a present process which is going on before our eyes. But when we reflect that the processes in the life of generations return periodically, we see that the conception of a particular stage of life as being in the future and operating at a distance is an arbitrary and hazardous conception, and that the stage of life in question can also be regarded as a process in the past, as something given which has left its traces behind. In this way the element of the unfamiliar and incomprehensible is greatly decreased. What we then have is not a possible future which *might* produce an effect, but a past which certainly *has* recurred countless times, and which certainly *has* produced an effect.

As an example of our contention that Physics is capable of co-operating fruitfully in the solution of what are apparently specifically biological problems, we have only to remember the remarkable progress made by experimental embryology and the mechanism of development with its physico-chemical methods. [7] We have another very remarkable fact in O. Wiener's demonstration of the probable connexion between color-photography and color-adaptation in nature. [8] By means of stationary light-waves stratification may be formed in a medium that is sensitive to light, and then the incident light may be reflected back as an interference-color; but there is yet another way in which a coloration corresponding to the illumination may arise. Of materials that are sensitive to light, there are some which can take on almost any hue. When such materials are exposed to a colored illumination, they retain the color of the illumination, because they do not absorb the rays of the same color as themselves, and consequently the light is incapable of producing any further change in them. According to Poulton's observations [9] it is probable that many of the adaptive colors of chrysalises arise in this manner. Thus in such cases we do not need to look far afield from the means that produce the effect, in order to find the "purpose" that is attained. Avoiding all rashness of statement, we may say that the equilibrium is determined by the circumstances under which it is attained.

8.

The conceptions of "effective cause" and of "purpose" both have their origin in animistic views, as is quite clearly seen from the scientific attitude of antiquity. The savage, no doubt, does not puzzle his head by reflecting on his own movements, which seem to him quite spontaneous, natural and self-evident. But as soon as he perceives unexpected and striking movements in nature, he instinctively interprets these movements on the analogy of his own. In this way the distinction between his own and someone else's volition begins to dawn upon him. [10] Gradually the similarities and differences between physical and biological processes standout alternately with ever greater clearness against the background of the fundamental scheme of volitional action. Where volitional action is conscious, cause and purpose still coincide. As regards physical processes, their great simplicity and their susceptibility to calculation cause the animistic conception to fade gradually away. Through a series of rigid forms the conception of cause merges by degrees into the conceptions of dependence and function. It is only for the phenomena of organic life, which offer less resistance to the animistic view, that the conception of purpose, the notion of an activity that is conscious of its end, is still maintained; and where conscious purposive action cannot be ascribed to the organism itself, some higher entity that strives towards a goal (Nature, or the like) is assumed as watching over the organism and guiding its activities.

Animism, or anthropomorphism, is not an epistomological fallacy; if it were, every analogy would be such a fallacy. The fallacy lies merely in the application of this view to cases in which the premises for it are lacking or are not sufficient. Nature, in producing man, has created a profusion of analogies between lower, and doubtless also between higher, stages of evolution.

When any process which is completely determined by the circumstances of the moment, and which remains limited to itself without further consequences, occurs in an inorganic, or even in an organic, body, we should scarcely speak of a purpose, as, for instance, when a sensation of light or a muscular contraction is excited by a stimulus. But when the hungry frog snaps at the fly which it sees, and swallows and digests it, we naturally adopt the notion of purposive action. Purposiveness only comes in when the organic functions are resolved into one another, when they are seen as interconnected, as not limited to the immediate, as proceeding by way of detours. In the sphere of the organic a much larger section of the world-process is manifested; we are aware of the influence of a wider spatial and temporal environment. That is why the organic is more difficult to understand. Real understanding is attained when, and only when, we have succeeded in resolving the complex into its immediately connected parts. Accordingly, the peculiar characteristics of the organic must be regarded only as provisional clues. The perusal of recent biological writings (Driesch, Reinke, etc.), though perhaps they are opposed to my own tendencies, only confirms me in this view. And if teleological investigation can only be provisional, the same is true of historical investigation also, since all historical research needs to be supplemented by causal explanation, a point which is very properly emphasized in Loeb's biological works, and in K. Menger's writings on economics.

9.

Every organism together with its parts is subject to the laws of physics. Hence the legitimate attempt gradually to conceive of an organism as something physical, and to establish the consideration of it in a "causal" point of view as alone valid. But whenever we try to do this we are always brought face to face with the peculiar characteristics of the organic, for which no analogy can be found in the physical phenomena of "lifeless" nature, so far as they have been investigated at present. Every organism is a system that is able to maintain its peculiar properties, its chemical composition, its temperature and so forth, in the face of external influences, and which manifests a state of dynamic equilibrium of considerable stability. [11] By an expenditure of energy the organism is able to draw more energy to itself from its environment, and thus to replace the loss of energy by an equal or a greater amount. [12] A steam-engine which should fetch its coal itself and heat itself, is only a feeble and artificial image of an organism. The organism possesses these properties even in its minuter parts; it regenerates itself from these parts; that is to say, it grows and propagates itself. Physics therefore still has much that is new to learn from a study of the organic before it is in a position to control the organic. [13]

The best physical image of a living process is still afforded by a conflagration, or some similar process, which automatically transfers itself to the environment. A conflagration keeps itself going, produces its own combustion-temperature, brings neighbouring bodies up to that temperature and thereby drags them into the process, assimilates and grows, expands and propagates itself. Nay, animal life itself is nothing but combustion in complicated circumstances. [14]

10.

Let us compare our volitional action with some reflex movement which we have observed in ourselves, and which causes us surprise when it occurs, or with the reflex movement of an animal. In the two latter cases we can detect an inclination to regard the whole process as physically determined by the momentary circumstances of the organism. Now what we call volition is nothing more than the totality of those conditions of a movement which enter partly into consciousness and are connected with a prevision of the result. If we analyse these conditions, so far as they enter into consciousness, we find nothing more than memory-traces of former experiences and their interconnection (association). It seems that the preservation of such traces and their associations is a fundamental function of elementary organisms, even though, in the case of such organisms, we are no longer able to speak of consciousness or of any arrangement in a system of memories.

If we may take memory and association, in Hering's wider sense, to be fundamental properties of elementary organisms, then adaptation would become intelligible. [15] Favourable combinations occur more often than in the ratio of compound probability, and remain associated. The presence of food, the feeling of satiety, and swallowing movements remain interconnected. The fact that phylogenesis is repeated in ontogenesis in an abbreviated form, would constitute a parallel to the well-known phenomenon by which thoughts return by preference along the paths which they have once taken, similar thoughts under similar conditions evoking similar thoughts. We do not indeed know what are the physical counterparts to memory and association. All the explanations that have been attempted are very much forced. In this respect it seems as if there were almost no analogy between the organic and the inorganic. It is possible, however, that, in the physiology of the senses, psychological observation on the one side and physical observation on the other, may make such progress that they will ultimately come into contact, and that in this way new facts may be brought to light. 1 The result of this investigation will not be a dualism, but rather a science which, embracing both the organic and the inorganic, shall interpret the facts that are common to the two departments.

[1] Cp. W. Pauli, *Physikalische-chemische Methoden in der Medizin,* Vienna, 1900, where an allied, but more narrowly limited, question is dealt with.

[2] *Prinzipien der Wärmelehre,* 2nd edition, Leipzig, 1900, pp. 434, 457.

[3] Cp. W. C. L. van Schaik, *Ueber die Tonerregung in Labialpfeifen.* Rotterdam, 1891; V. Hensen, *Annalen der Physik,* 4th Series, Vol. III., p. 719, 1900.

[4] *Die Geschichte und die Wurzel des Satzes der Erhaltung der Arbeit,* Prague, Calve, 1872. (English translation, *History and Root of the Principle of the Conservation of Energy,* by P. E. B. Jourdain, Chicago, Open Court Publishing Co., 1911.)

[5] Such objections have been raised by Külpe in his *Ueber die Beziehungen zwischen körperlichen und seelischen Vorgängen* (*Zeitschrift für Hypnotismus,* Vol. VII., p. 97), also by Cossman in his *Empirische Teleologie,* Stuttgart, 1899, p. 22. I do not think that my view differs so greatly from Cossman's that an understanding is impossible. If he had considered the matter further, he would have seen that I substituted the notion of function for the old notion of causality, and that the notion of function is sufficient also for those cases which he has in view. I have no further objection to make to his "Empirical teleology." Cp. also C. Hauptmann *Die Metaphysik in der Physiologie,* Dresden, 1893.

[6] Cp. *Erkenntnis und Irrtum* (1905), p. 274.

[7] Cp. W. Roux, *Vorträge und Aufsätze über Entwickelungsmechanik der Organismen,* Leipzig, W. Engelmann, 1905.

[8] O. Wiener, "Farbenphotographie und Farbenanpassung in der Natur," Wiedemann's *Annalen,* Vol. LV., 1895, p. 225.

[9] Poulton, *The Colors of Animals,* London, 1890.

[10] I once set a Holtz electrical machine going for the benefit of one of my boys when he was about three years old. He was delighted by the dancing sparks. But when I let go the machine and it went on rotating by itself, he started back in terror and apparently thought it was alive: "It goes by itself!" he exclaimed, in startled and anxious tones. Perhaps it is the same with dogs when they run barking after every moving cart. (For another plausible explanation, which does not agree with this view, see Zell, *Sind die Tiere unvernünftig?* Kosmosverlag, p. 38.) I remember that when I was about three years old I was frightened when the elastic seed-capsule of a plant of gardenbalsam opened on being pressed, and pinched my finger. It seemed to me alive, like an animal.

[11] Hering, *Vorgänge in der lebendigen Substanz,* Lotos, Prague, 1888.

[12] Hirth, *Energetische Epigenesis,* Munich, 1898, pp. x., xi.

[13] Hering, *Zur Theorie der Nerventätigkeit,* Leipzig, 1899.

[14] Cf. Ostwald, *Naturphilosophie,* and the work of Roux cited on p. 95, above.

[15] Hering, *Ueber das Gedächtnis als allegemeine Funktion der organisierten Materie,* Vienna, 1870.

[16] I first tentatively suggested this notion, though still in terms of Fechner's theories, in the *Kompendium der Physik für Mediziner,* 1863, p. 234.

VI. The Space Sensations of the Eye

1.

A tree with its hard, rough, grey trunk, its many branches swayed by the wind, its smooth, soft, shining leaves, appears to us at first a single, indivisible whole. In like manner, we regard the sweet, round, yellow fruit, the warm, bright fire, with its manifold moving tongues, as a single thing. One name designates the whole, one word draws forth from the depths of oblivion all the associated memories at once, as if they were strung upon a single thread.

The reflexion of the tree, the fruit, or the fire in a mirror is visible, but not tangible. When we turn our glance away or close our eyes, we can touch the tree, taste the fruit, feel the fire, but we cannot see them. Thus the apparently indivisible thing separates into parts, which are not only attached to one another but also to other conditions. The visible is separable from the tangible, from that which may be tasted, etc.

What is merely visible also appears at first sight to be a single thing. But we may see a round, yellow fruit together with a yellow, star-shaped blossom. A second fruit may be just as round as the first, but is green or red. Two things may be alike in color but unlike in form; they may be different in color but like in form. Thus sensations of sight are separable into color-sensations and space -sensations, which are different from one another even though they cannot be represented in isolation from another.

2.

Color-sensation, into the details of which we shall not enter here, is essentially a sensation of favorable or unfavorable chemical conditions of life. In the process of adaptation to these conditions, color-sensation has probably been developed and modified. [1] Light introduces organic life. The green chlorophyll and the (complementary) red haemoglobin play a prominent part in the chemical processes of the plant-body and in the chemical reactions of the animal body. The two substances present themselves to us in the most varied modifications of tint. The discovery of the visual purple, and observations in photography and photo-chemistry, allow us to conceive visual processes also as chemical processes. The role which color plays in analytical chemistry, in spectrum-analysis, in crystallography, is well known. It suggests a new conception for the so-called vibrations of light, according to which they should be regarded, not as mechanical, but as chemical vibrations, as successive union and separation, as an oscillatory process of the same sort that takes place, though only in one direction, in photo-chemical phenomena. This conception, which is substantially supported by recent investigations in anomalous dispersion, accords with the electro-magnetic theory of light. In the case of electrolysis, in fact, chemistry yields the most intelligible conception of the electric current by regarding the two components of the electrolyte as passing through each other in opposite directions. It is likely, therefore, that in a future theory of colors, many biologico-psychological and chemico-physical threads will be united.

3.

Adaptation to the chemical conditions of life which manifest themselves in color, renders locomotion necessary to a far greater extent than adaptation to those which manifest themselves through taste and smell. At least this is so in the case of man, which is here in question, and as to which alone a direct and certain judgment is within our power. The close association of space-sensation (a mechanical factor) with color-sensation (a chemical factor) is thus rendered intelligible. We shall now proceed to the analysis of optical space-sensations.

4.

In examining two figures which are alike but differently colored (for example, two letters of the same size and shape, but of different colors), we recognize their sameness of form at the first glance, in spite of the difference of color-sensation. The sight-perceptions, therefore, must contain some identical sensation-components. These are the space-sensations which are the same in the two cases.

Fig. 2.

5.

We will now investigate the character of the space-sensations that physiologically condition the recognition of a figure. First, it is clear that this recognition is not the result of geometrical considerations which are a matter, not of sensation, but of intellect. On the contrary, the space-sensations in question serve as the starting-point and foundation of all geometry. Two figures may be geometrically congruent, but physiologically quite different, as is

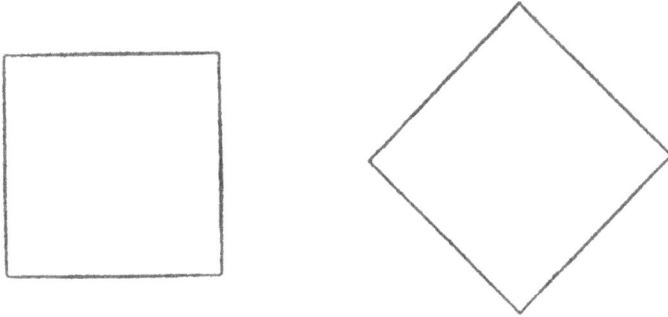

Fig. 3.

Fig. 4.

shown by the two adjoined squares (Fig. 3), which could never be recognized as the same without mechanical and intellectual operations. [2] A few simple experiments will render us famil-iar with the relations here

Fig. 5.

involved. Look at the spot in Fig. 4. Place the same spot twice or several times in exactly the same position in a row (Fig. 5); the result is a peculiar, agreeable impression, and we recognize at once and without difficulty the identity of all the figures. When, however, we turn one spot far enough round with respect to the other (Fig. 6), their identity of form is not recognizable without intellectual assistance. On the other hand, if we place two of the spots in positions symmetrical to the median plane of the observer (Fig. 7) the relationship of form is strikingly apparent. But if the plane of symmetry diverges considerably from the median plane of the observer, as in Fig. 8, the affinity of form is recognizable only by turning the figure around or by an intellectual act. On the other hand, the affinity of form is again apparent on contrasting with such a spot the same spot rotated through an angle of 180° in the same plane (Fig. 9). In this case we have the so-called centric symmetry.

If we reduce all the dimensions of the spot proportionately, we obtain a geometrically similar spot. But as the geometrically

Fig. 6.

Fig. 7.

Fig. 8.

congruent is not necessarily physiologically (optically) congruent, nor the geometrically symmetrical necessarily optically symmetrical, analogously the geometrically similar is not necessarily optically similar. It is only when the two geometrically similar spots are placed beside each other in the same relative positions (Fig. 10), that they will also appear optically similar. Turning one of the spots round destroys the resemblance (Fig. 11). If we substitute for one of the spots a spot symmetrical to the other in respect of the median plane of the observer (Fig. 12), a symmetrical similarity will be produced which has also an optical value. The turning of one of the figures through 180° in its own plane, producing thereby centrically symmetrical similarity, has also a physiologico-optical value (Fig. 13).

Fig. 9.

6.

In what, now, does the essential nature of optical similarity, as contrasted with geometrical similarity, consist? In geometrically similar figures, all homologous distances are proportional. But that is an affair of the intellect, not of sensation. If we place beside a triangle with the sides a, b, c, a triangle with the sides $2a$, $2b$, $2c$, we do not recognize this simple relation between the two immediately, but intellectually, by measurement. If the similarity is to become optically perceptible, the proper position must be added. That a simple relation of two objects for the intellect does not necessarily condition a similarity of sensation, may be perceived by comparing two triangles having respectively the sides, a, b, c, and $a + m$, $b + m$, $c + m$. The two triangles do not look at all alike. Similarly all conic sections do not look alike, although all stand in a simple geometric relation to each other; still less do curves of the third order exhibit optical similarity, etc.

Fig. 10.

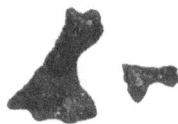

Fig. 11.

7.

The geometrical similarity of two figures is determined by all their homologous lines being proportional or by all their homologous angles being equal. But to appear optically similar the figures must also be *similarly situated,* that is all their homologous directions

Fig. 12

Fig. 13.

must be parallel or, as we prefer to say, must be the same (Fig. 14). The importance of direction for sensation will be evident upon a careful consideration of Fig. 3. It is by identity of direction, accordingly, that are determined the identical space-sensations which are characteristic of the physiologico-optical similarity of the figures. [1]

We may obtain an idea of the physiological significance of the direction of a given straight line or curve-element, by the following reflexion. Let $y = f(x)$ be the equation of a plane curve. We can read at a glance the course of the values of dy/dx on the curve, for they are determined by its steepness; and the eye gives us, likewise, qualitative in-

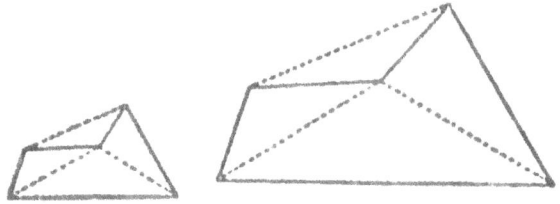

Fig. 14

formation concerning the values of d_2y/dx_2, for they are characterized by the curvature. The question naturally presents itself, why can we not arrive at as immediate conclusions concerning the values d_3y/dx_3, d_4y/dx_4, etc. The answer is easy. What we see is of course not the differential coefficients, which are an intellectual affair, but only the direction of the curve-elements, and the deviation of the direction of one curve-element from that of another.

In fine, since we are immediately cognisant of the similarity of figures lying in similar positions, and are also able to distinguish at once the special case of congruity, therefore our space-sensations yield us information concerning identity or difference of directions and equality or inequality of dimensions.

8.

It is *a priori* extremely probable that sensations of space are connected in some way with the motor apparatus of the eye. Without entering into particulars, we may observe, first, that the whole apparatus of the eye, and especially the motor apparatus, is symmetrical with respect to the median plane of the head. Hence, symmetrical movements of looking will be connected with like or approximately like space-sensations. Children constantly confound the letters b and d, p and q. Adults, too, do not readily notice a change from left to right, unless some special points of apprehension for sense or intellect make it noticeable. The symmetry of the motor apparatus of the eye is very perfect. The like excitation of its symmetrical organs would, by itself, scarcely account for the distinction of right and left. But the whole human body, especially the brain, is affected with a slight asymmetry, which leads, for example, to the preference of one (generally the right) hand, in motor functions. And this leads, again, to a further and better development of the motor functions of the right side, and to a modification of the attendant sensations. After the space-sensations of the eye have become associated, through writing, with the motor sensations of the right hand, a confusion of those vertically symmetrical figures with which the art and habit of writing are concerned no longer ensues. This association may even become so strong that the memories follow only the accustomed tracks, and we read, for example, the reflexion of written words in a mirror only with the greatest difficulty. The confusion of right and left still occurs, however, with regard to figures which have no motor, but only a purely optical (for example, ornamental) interest. A noticeable difference between right and left must be felt, moreover, by animals, as in many predicaments they have no other means of finding their way. How similar, moreover, are the sensations connected with symmetrical motor functions is easily remarked by the attentive observer. If, for example, because my right hand happens to be engaged, I grasp a micrometer-screw or a key with my left hand, I am certain (unless I reflect beforehand) to turn it in the wrong direction, - that is, I always perform the movement which is symmetrical to the usual movement, confusing the two because of the similarity of the sensation. The observations of Heidenhain regarding the reflected writing of persons hypnotized on one side should also be cited in this connexion.

9.

The idea that the distinction between right and left depends upon an asymmetry, and possibly in the last resort upon a chemical difference, is one which has been present to me from my earliest years. I gave expression to it in the first lectures I ever delivered, in 1861. Since then this idea has forced itself upon me again and again. I learned by chance from a retired army-officer that on dark nights or in snow-storms, when external landmarks are absent, troops will move approximately in a circle of large radius so that they almost return to their point of departure, though all the time they are under the impression that they are marching straight forward. An analogous phenomenon is narrated in Tolstoi's story, *Master and Servant*. Probably the only way to understand these phenomena is to assume a slight motor asymmetry. They are analogous to the way in which a ball with a slight deviation from the true cylindrical shape rolls in a circle of large radius. This is actually the way in which the matter is regarded by F. O. Guldberg, [4] who has carried out detailed researches on the phenomena presented in this connexion by human beings and animals that have lost their way. Human beings and animals that have lost their direction move, almost without exception, nearly in circles, of which the radii vary according to the species, while the centre lies sometimes on the left hand of the individual travelling along the circumference, and sometimes on his right, according to the individual and the species. According to Guldberg we have here a teleological device to help parents to find their hungry young again when they have been lost. Experiments on the lower animals, with whom this factor is absent, would therefore be interesting. For the fest, we should expect, on grounds of general probability, to find imperfect symmetry in the lower animals also.

Again, Loeb's researches "On the Spatial Feeling of the Hand," [5] have taught us, amongst other things, that when the eyes are bandaged, a given movement of the right hand, if imitated by the left, is always reproduced in an exaggerated or a diminished form; the degree of exaggeration or diminution varying with the individual. Loeb thinks that the phenomena of regeneration allow us to infer that the distinction between right and left is specific. I am certain, however, that I personally have never regarded it as a merely geometrical and quantitative motor difference.

10.

With looking upwards and looking downwards, fundamentally different space sensations are associated, as ordinary experience shows. This is, moreover, comprehensible, since the motor apparatus of the eye is asymmetrical with respect to a horizontal plane. The direction of gravity also is so very decisive and important for the motor apparatus of the rest of the body that this fact has assuredly also found its expression in the apparatus of the eye, which serves the rest. It is well known that the symmetry of a landscape and of its reflexion in water is not felt. The portrait of a familiar personage, when turned upside down, is strange aid puzzling to a person who does not recognize it intellectually. If we place ourselves behind the head of a person lying upon a couch, and unreflectingly give ourselves up to the impression which the face makes upon us, we shall find that our impression is altogether strange, especially when the person speaks. The letters *b* and *p*, and *d* and *q*, are not confused even by children.

Our previous remarks concerning symmetry, similarity, and the rest, naturally apply not only to plane figures, but also to those in space. Hence, we have yet a remark to add concerning the sensation of space-depth. The sight of something at a distance causes different sensations from the sight of something near at hand. These sensations must not be confused, because of the supreme importance of the difference between near and far, both for animals and human beings. They *cannot* be confused, because the motor apparatus is asymmetrical with respect to a plane perpendicular to the direction from front to rear. It is a common experience that a portrait-bust of a person whom we know quite well cannot be replaced by the mould in which the bust is cast, and this experience is quite analogous to the observations consequent upon the inversion of objects.

11.

If we suppose that identical dimensions and identical directions excite identical space-sensations, and that directions symmetrical with respect to the median plane of the head excite similar space-sensations, it becomes easy to understand the above-mentioned facts. The straight line has, in all its elements, the same direction, and everywhere excites the same space-sensations. Herein consists its aesthetic value. Moreover, straight lines which lie in the median plane or are perpendicular to it are brought into special relief by the circumstance that, thanks to this position of symmetry, they stand in the same relation to both of the two halves of the visual apparatus. Every other position of the straight line is felt as an obliquity, as a deviation from the position of symmetry.

The repetition of the same space-figure in the same position conditions a repetition of the same space-sensation. All lines connecting prominent (noticeable) homologous points have the same direction and excite the same sensation. Likewise when merely geometrically similar figures are placed side by side in the same positions, this relation holds. The sameness of the dimensions alone is absent. But when the positions are disturbed, this relation, and with it, the impression of unity the aesthetic impression are also disturbed.

In a figure symmetrical with respect to the median plane, *similar* space-sensations corresponding to the symmetrical directions take the place of the *identical* space-sensations. The right half of the figure stands in the same relation to the right half of the visual apparatus as the left half of the figure does to the left half of the visual apparatus. If we alter the sameness of the dimensions, the sensation of symmetrical similarity is still felt. An oblique position of the plane of symmetry upsets the whole relation.

If we turn a figure through 180°, contrasting it with itself in its original position, centric symmetry is produced. That is, if two pairs of homologous points be connected, the connecting lines will cut each other at a point *O*, through which, as their point of bisection, all lines connecting homologous points will pass. Moreover, in the case of centric symmetry, all lines of connexion between homologous points have the same direction, - a fact which produces an agreeable sensation. If the sameness of the dimensions is eliminated, there still remains, for sensation, centrically symmetrical similarity.

Regularity appears to have no special physiological value, in distinction from symmetry. The value of regularity probably lies only in its *manifold* symmetry, which is perceptible in more than one *single* position.

12.

The correctness of these observations will be apparent on glancing over the work of Owen Jones - *A Grammar of Ornament,* London, 1865. In almost every plate one finds new and different kinds of symmetry as fresh testimony in favor of the conceptions above advanced. The art of decoration, which, like pure instrumental music, aims at no ulterior end, but ministers only to pleasure in form and color, is the best source of material for our present studies. Writing is governed by other considerations than that of beauty. Nevertheless, we find among the twenty-four large Latin letters ten which are vertically symmetrical (A, H, I, M, O, T, V, W, X, Y), five which are horizontally symmetrical (B, C, D, E, K), three which are centrically symmetrical (N, S, Z), and only six which are unsymmetrical (F, G, L, P, Q, R).

The study of the evolution of primitive art is extremely instructive in connexion with the problems under discussion. The character of primitive art is determined, firstly, by the natural objects that offer themselves to the imitation of the artist j secondly, by the degree of mechanical skill attained; and finally by the effort to make use of repetition in its various forms. [6]

13.

I have clearly explained briefly in previous writings the aesthetic significance of the above-mentioned facts, to treat of which in detail was not part of my plan. I cannot, however, refrain from mentioning that this has been done by a physicist, the late J. L. Soret of Geneva, in an admirable book published in 1892, [7] for which the way was prepared by a lecture delivered by him at the meeting of the Swiss Scientific Association in 1866. Soret's views are connected with those of Helmholtz, and he does not seem to be acquainted with my theories. He does not go into the physiological side of the question, but on the aesthetic side his exposition is very copious and illustrated by appropriate examples. He discusses the aesthetic effect of symmetry, of repetition, of similarity and of continuity, which last he regards as a special case of repetition. According to him slight deviations from symmetry can more than compensate for the loss in sensual satisfaction by the multiplicity which they introduce and by the intellectual aesthetic pleasure bound up with that multiplicity. This is illustrated from the sculptured ornaments of Gothic cathedrals. This intellectual pleasure is also produced by the virtual or potential symmetry which we perceive when the human figure or some other symmetrical form is placed in an asymmetrical position. And he does not merely apply these reflexions to optical cases, but extends them to all departments, as also has been done by me. He notices rhythm, music, movements, dancing, the beauties of Nature, and even literature. Particularly interesting are his observations on blind people, which the Asylum for the Blind at Lausanne gave him the opportunity of carrying out. Blind people take pleasure in the periodic repetition of the same forms in tangible objects, and have a decided sense of formal symmetry. Striking disturbances in symmetry of form are unpleasant to them, and sometimes even seem to them ludicrous. A blind man, who had been accustomed to study a large-scale map of Europe in relief, recognized that continent by means of geometrical similarity when he found it as part of a larger raised map on a smaller scale. The symmetrical organs of touch, the two arms and hands, are in fact arranged in an analogous way to the organs of sight, so that we need not be surprised at the agreement. Even in antiquity this agreement was not without its influence upon inquirers, to say nothing of a modern thinker like Descartes; it even produced a number of unfortunate notions which are partly operative even to-day. Soret's chapter on literature seems less successful. True, in metre, rhyme, etc., we have phenomena similar to those noticed in the previously treated departments. But when, for example, he draws a parallel between the effect of the sixfold repetition of the phrase "Que diable allait il faire dans cette galère," in Molière's well-known play, [8] and the repetition of an ornamental motive, probably few will agree with him. It is certain that the effect of this repetition is not produced by the repetition as such, but by the successive heightening of a comic contrast, and that consequently it is merely intellectual.

Finally, I should like to draw the attention of the reader to an article by Arnold Emch in *The Monist* for October, 1900, on "Mathematical Principles of Aesthetic Forms." Emch gives attractive examples of the way in which a number of forms arranged in a series co-operate to produce an aesthetic impression by observing one and the same geometrical principle. He is following out the line of thought on which I touched in my lecture of 1871, according to which production in accordance with a fixed rule has an aesthetic effect (*Popular Scientific Lectures*, Chicago, Open Court Publishing Co., 1894). But in that place I laid stress upon the point, and should like to do so again, that the rule, considered as an affair of the intellect, has no aesthetic effect in itself, but that the effect depends on the repetition, determined by the rule, of one and the same sensational motive.

14.

Here we must once more point out that the geometrical and. the physiological properties of a figure in space are to be sharply distinguished. The physiological properties are determined by the geometrical properties coincidently with these, but are not determined by these solely. On the other hand, physiological properties very probably gave the first impulse to geometrical investigations. The straight line doubtless first attracted attention not because of its being the shortest line between two points, but because of its physiological simplicity. The plane likewise possesses, in addition to its geometrical properties, a special physiologico-optical (aesthetic) value, which causes it to be noticed, as will be shown later on. The division of the plane and of space by right angles has not only the advantage of producing equal parts, but also an additional and special symmetry-value. The circumstance that congruent and similar geometrical figures can be brought into positions where their relationship is physiologically felt, led, no doubt, to an earlier investigation of these kinds of geometrical relationship than of those that are less noticeable, such as affinity, collineation, and others. Without the co-operation of sense-perception and understanding, 'a scientific geometry is inconceivable. But H. Hankel has admirably shown in his *History of Mathematics* (Leipzig, 1874) that in Greek geometry the factor of pure understanding is decidedly dominant, whereas in Indian geometry the factor of sense has the upper hand. The Hindus make use of the principles of symmetry and similarity (see, for example, p. 206 of Hankel's book) with a generality which is totally foreign to the Greeks. Hankel's proposal to unite the rigor of the Greek method with the perspicuity of the Indian in a new mode of presentation is well worthy of encouragement. Furthermore, in so doing, we should only be following in the footsteps of Newton and John Bernoulli, who, even in mechanics, applied the principle of similarity in a still more general manner. The advantages that the principle of symmetry affords in the last-named department, I have shown at length elsewhere. [9]

[1] Compare Grant Allen, *The Color-Sense,* London, Trübner & Co., 1879. The attempt of H. Magnus to show a considerable development of the color-sense within historical times, cannot, I think, be regarded as felicitous. Immediately after the appearance of the writings of Magnus, I corresponded with a philologist, Prof. F. Polle of Dresden, on this subject, and both of us soon came to the conclusion that the views of Magnus could not hold their own before the critical examination either of natural science or of philology. As each of us left the publication of the results of our discussion to the other, these were never made public. Meantime, however, the matter has been disposed of by E. Krause, and in detail by A. Marty. I shall take the liberty of adding only a few brief remarks. From defects of terminology we cannot infer the absence of corresponding qualities of sensation. Terms, even to-day, are always indistinct, hazy, defective, and few in number, where there is no necessity for sharp discrimination. The color-terminology of the countryman of to-day, and his terminology of sensations in general, is no more developed than that of the Greek poets. The peasants of the Marchfeld say, for example, as I have often proved by personal experience, that salt is "sour," because the expression "salty" is not familiar to them. The terminology of colors must not be looked for in the poets, but in technical works. And, furthermore, as my colleague Benndorf has remarked, we must not take an enumeration of vase -pigments for an enumeration of all colors, as does Magnus. When we consider the polychromy of the ancient Egyptians and Pompeiians, when we take into account the fact that these decorations can scarcely have been produced by the color-blind, when we note that Pompeii was buried in ashes only seventy years after Vergil's death, whilst Vergil on this theory is supposed to have been nearly color-blind, the untenability of the whole conception is sufficiently apparent. The question has lately been taken up again, with recourse to fuller authorities, by W. Schultz (*Das Farbenempfindungssystem der Hellenen,* Leipzig, 1904). Applications of the Darwinian theory are also to be made with caution in another direction. We like to picture to ourselves a condition in which the color-sense is lacking, or in which little color-sense exists, as preceding another in which the color-sense is highly developed. For the beginner it is natural to proceed from the simple to the complex. But this is not necessarily the path of Nature. The color-sense exists, and it is probably variable. But whether it is being enriched or impoverished who can tell? Is it not possible that, with the awakening of intelligence and the use of artificial contrivance, the whole development will be shifted to the intellect, which certainly is chiefly called into play from this point on, and that the development of the lower organs of man will be relegated to second place?

[2] Compare my brief paper, *Ueber das Sehen von Lagen und Winkeln,* in the *Sitzungsberichte der Wiener Akademie,* Vol. XLIII., 1861, p. 215.

[3] Some forty years ago, in a society of physicists and physiologists, I proposed for discussion the question, why geometrically similar figures were also optically similar. I remember quite well the attitude taken with regard to this question, which was accounted not only superfluous, but even ludicrous. Nevertheless, I am now as strongly convinced as I was then that this question involves the whole problem of form-vision. That a problem cannot be solved which is not recognized as such is clear. In this non-recognition, however, is manifested, in my opinion, that one-sided mathematico-physical direction of thought, which alone accounts for the opposition, from so many sides and extending over so many years, instead of cheerful acceptance, with which the writings of Hering have been received.

[4] F. O. Guldberg, "Die Zirkularbewegung," *Zeitschrift für Biologie,* Vol. XXV., p. 419, 1897. Dr W. Pauli drew my attention to this article in conversation.

[5] Loeb, "Ueber den Ftihlraum der Hand," Pflüger's *Archiv,* Vols, XLI. and XLVI.

[6] Alfred C. Haddon, *Evolution in Art, as illustrated by the Life-histories of Designs;* London, 1895.

[7] J. L. Soret, *Sur les conditions physiques de la perception du beau,* Geneva, 1892.

[8] *Les Fourberies de Scapin.*

[9] I have given less complete discussions of the leading thoughts of this chapter in the paper already mentioned, *Ueber das Sehen von Lagen und Winkeln* (1861), further (in Fichte's *Zeitschrift für Philosophie,* Vol. XLVI., 1865, p. 5, and in *The Forms of Liquids, and Symmetry* (1872) now also published in my *Popular Scientific Lectures,* translated by Thomas J. McCormack, Open Court Publishing Co., Chicago, 1894. With regard to the use of the principle of symmetry in mechanics, compare my work *The Science of Mechanics* (1883), translated by Thomas J. McCormack, 1893, Open Court Pub. Co., Chicago.

VII. Further Investigation of Space-Sensations

[1]

1.

OUR knowledge of spatial vision made important advances in the course of the nineteenth century, not merely because a gain in positive understanding was involved, but also because the prejudices accumulated by various philosophers and physicists in this department, especially since Descartes, have been finally disposed of, and thereby that freedom from preconceptions attained which is the first requisite for making positive discoveries.

Johannes Müller, [2] created the doctrine of specific energies, and also put forward with great lucidity the theory of identical retinal positions, which, for the rest, can be clearly traced back as far as Ptolemy. [3] On Müller's theory that the retina has sensations of itself in its own activity, "visual space" is, for him, something immediately given. My own body also appears in my field of vision. All questions of direction can only refer to the relative positions of parts of the field of vision. The direction of vision depends exclusively on the arrangement of the sensitive parts of the retina. All theories as to projection, and problems as to why we see things upright, disappear. But estimation of the distance of an object seen is, for Müller, still through and through an affair of the intellect.

Wheatstone's [4] discovery of the spectroscope led at once to the conviction that in certain circumstances images could be seen as simple, and with different degrees of depth according to the stereoscopic difference, not only when the images fell upon identical parts of the retina, but also when they fell upon other parts, provided the difference between the parts was not too great. The result of this was to throw doubt on the doctrine of identity, and to stimulate the formulation of psychological explanations of how we come to see things as having depth. Hence arose Brücke's theory of successive fixation in spatial vision, which in its turn was proved to be untenable by Dove's experiments in instantaneous illumination with the stereoscope.

Panum [5] opposed these theories with arguments of great force, and by admirably contrived experiments. Taking his stand on the phenomenon of binocular rivalry and the prominent part played therein by contours, he came to the conclusion that our seeing things as having depth depends upon a reciprocal action, or "Synergy," of the two retinae, and that the sensation of depth is an innate specific energy. The more similar the two monocular images, and especially the contours, are in form, color, and position, the more easily do they coalesce into a single binocular image, of which the depth is determined by the stereoscopic difference. But Panum still maintains that this depth corresponds to what is given by means of lines of projection.

It is to Hering [6] that we owe the most thorough clearing away of old prejudices. His starting-point is the view that our immediately given visual space must be completely distinguished from the conceptual space which we obtain by means of experiences of a special kind. He proves by decisive experiments that the direction in which we see an object is different from the line, the line of vision, or of projection, between the object and the retinal image. There are two lines of vision, one for each eye; but there is only one direction of vision, bisecting the angle formed by these two lines. We have to think of this direction of vision as proceeding from the point of bisection of the line connecting the two eyes. In order to exclude all reference to geometrical space, we may put it thus: The two eyes together see the same relative horizontal and perpendicular arrangements that a single eye would see if it were situated half-way between the two eyes. Suppose that, looking in a horizontal direction and with symmetrical convergence, we fix our gaze upon a point on the windowpane; then we see this point in the median plane, but at the same time we see behind it in the same plane objects lying a long way to one side. In the stereoptical experiment we still see objects in front of us, even when there is only a slight divergence of the axes of the eyes, although the directions of projection no longer lead to these objects, or at any rate have no longer any physical or physiological meaning. Again, the distances which we see do not agree with the results of the theory of projection. When, with horizontal lines of vision, we stretch vertical threads through Müller's circular horopter, the cylinder thereby produced appears to us like a plane. We see not only the image of the fixed or "nucleus" point, but also the collective conception of all the points (the "nucleus-surface") represented in identical, or "corresponding," positions, as a plane lying before us at a definite distance. It is impossible to explain these and many other analogous facts on the theory of projection. Hering reduces spatial vision to a simple principle. Identical, or "corresponding," points on the retina have identical height and breadth-values; symmetrical points on the retina, on the other hand, have identical depth-values, and these last increase from the outer edges of the retina inwards. When similarity of the monocular images in color, shape and position causes them to coalesce into a singular binocular image, the binocular image contains the mean value of the depth-values of the single images. Such mean values of the single images play a decisive part in general, and in particular as regards the directions of vision. These indications must be sufficient for us here, since we cannot now discuss in detail the varied contents of the works in which Hering has laid the secure foundations on which this chapter is based. [7] The only further remark I will make is, that, according to him, the two eyes are to be conceived as a single united organ, the associated movements of which rest on an innate anatomical foundation, - a point that had already been brought out by Johannes Müller.

Biological and psychological [8] research combine to confirm the conclusion that, as regards the intuition of space, the nativistic view can *a fortiori* be maintained. The chick has scarcely broken from its shell than it is seen to be at home in space and pecking at everything that excites its attention. For the new-born human being we can at most suppose only a lower degree of maturity, but otherwise the conditions must be essentially the same. Panum has brought out this point. Spatial intuition is therefore present at birth. Whether we shall ever be in a position to explain it, in the sort of way attempted by Helmholtz, by means of the history of evolution or the history of the race, is a separate question.

Perhaps clues towards the solution of this problem may be found in the facts of phylogenetic development and the variation of retinal correspondence (investigated by Johannes Müller) [9] which takes place at the transition between one animal species and another. Another promising field for research is presented by the pathological anomalies of people who squint and the phenomena of adaptation to be observed in these cases. [10]

2.

That space-sensation is connected with motor processes has long since ceased to be disputed. Opinions differ only as to how this connexion is to be understood.

If two congruent images of different colors fall in succession on the same parts of the retina, they are at once recognized as identical figures. We may, therefore, regard different space-sensations as connected with different parts of the retina. But that these space-sensations are not unalterably con-

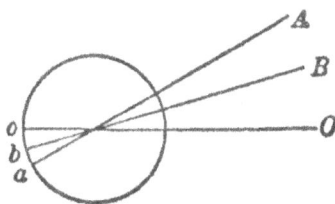

Fig. 15.

nected with particular parts of the retina, we perceive on moving our eyes

77

freely and voluntarily, whereby the objects observed do not change their position or form, although their images are displaced on the retina.

If I look straight before me, fixing my eyes upon an object O, an object A, which is reflected on the retina in a, at a certain distance below the point of most distinct vision, appears to me to be situated at a certain height. If I now raise my eyes, fixing them upon B, A retains its former height. It would necessarily appear lower down if the position of the image on the retina, or the arc oa, alone determined the space-sensation. I can raise my glance as far as A and farther without any change in this relation. Thus, the physiological process which conditions the *voluntary* raising of the eye, can entirely or partly take the place of the height-sensation, is homogeneous with it, or, in brief, algebraically summable with it. If I turn my eyeball upward by a slight pressure of the finger, the object A actually appears to sink, proportionately to the shortening of the arc oa. The same thing happens when, by any other unconscious or involuntary process for example, through a cramp of the muscles of the eye the eyeball is turned upward. According to an experience now familiar to oculists for some decades patients with paralysis of the rectus externus reach too far to the right in attempting to grasp objects at the right. Since they need to exert a stronger impulse of the will than persons of sound eyes, in order to fix their glance upon an object to the right, the thought naturally suggests itself that the will to look to the right determines the optical space-sensation "right." Some years ago, [11] I put this observation into the form of an experiment, which every one can try for himself. Let the eyes be turned as Jar as possible towards the left and two large lumps of moderately hard putty firmly pressed against the right side of each eye-ball. If, now, we attempt to glance quickly to the right, we shall succeed only very imperfectly, owing to the imperfectly spherical form of the eyes, and the objects will suffer a strong displacement to the right. Thus the mere will to look to the right imparts to the images at certain points of the retina a larger "rightward value," as we may term it for brevity. The experiment is, at first, surprising. It will soon be perceived, however, that both facts - viz., that by voluntarily turning the eyes to the right, objects are not displaced, and that by the forced, involuntary turning of the eyes to the left, objects are displaced to the right - together teach the same lesson. My eye, which I wish to, and cannot, turn to the right, may be regarded as voluntarily turned to the right and compulsorily turned back by an external force. Professor W. James [12] could not get the experiment just mentioned to succeed. I have often repeated it, and always found it confirmed. I think that the fact is certain, but of course that decides nothing as to its correct interpretation.

3.

The will to perform movements of the eyes, or the innervation to the act, is itself the space-sensation. This follows naturally from the preceding consideration. [13] If we have a sensation of itching or pricking in a certain spot of our skin, by which our attention is sufficiently secured, we immediately grasp at the spot with the correct amount of movement. In the same manner we turn our eyes with the correct amount of exertion towards an object reflected on the retina, as soon as this exerts a sufficient stimulus to draw our attention. By virtue of organic arrangements and long exercise we hit immediately upon the exact degree of innervation sufficient to enable us to fix our eyes upon an object reflected on a certain point of the retina. If the eyes are already turned towards the right, and we begin to give our attention to an object further to the right or the left, a new innervation of the same sort is algebraically added to that already present. A disturbance of the process arises only when alien, involuntary innervations or externally moving forces are added to the innervations determined by the will.

4.

Years ago, while occupied with the questions now under discussion, I noticed a peculiar phenomenon, which has not yet, so far as I know, been described. In a very dark room we fix our eyes upon a light A, and then suddenly look at a light lower down, B. At this, the light A appears to make a rapid sweep A A' (quickly ended)

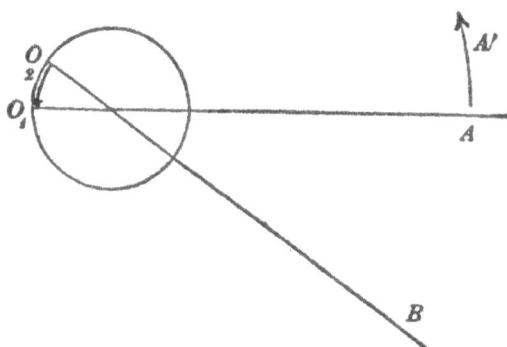

Fig. 16.

upwards. The light B, of course, does the same; but to avoid complications, this is not indicated in the diagram. The sweep is, of course, an after-image, which enters consciousness only on completion, or shortly before completion, of the glance-movement, but - and this is the remarkable point - with positional values that correspond, not to the later, but to the earlier innervations and position of the eye. Similar phenomena are often noticed in experiments with Holtz's electrical machine. If the experimenter is surprised by a spark during a glance downwards, the spark often appears high above the electrodes. If it yields a lasting after-image, the image appears, of course, below the electrodes. The preceding phenomena answer to the so-called personal equation of astronomers, except that they are confined to

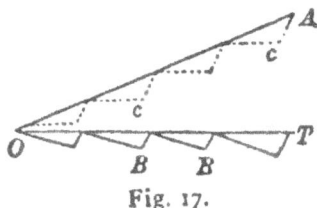

Fig. 17.

the province of sight. By what organic arrangements this relation is determined must be left an open question, but it is probably of some value in preventing confusion of position in movements of the eyes. [14]

5.

For the sake of simplicity we have hitherto regarded only the eyes as in motion, and have considered the head and the body generally as at rest. If, now, we move the head about without intentionally fixing the eyes upon any object, the objects seen remain motionless. But at the same time another observer may notice that the eyes, like frictionless, inert masses, take no part in the turning movements. Still more noticeable is the phenomenon if we turn with continuous motion, actively or passively, about a vertical axis, as viewed from above, say, in the direction of the hands of a clock. In this case, as Breuer has observed, the open or closed eyes turn, about ten times to a full revolution of the body, in the opposite direction to that of the clock-hands, with a uniform motion, and as frequently back again in the opposite direction by jerks. The process is represented in the diagram of Fig. 17. On $O\,T$, the times are laid off as abscissae, the angles described in the direction of the clock-hands are laid off as ordinates upwards, and the angles described in the opposite direction as ordinates downwards. The curve $O\,A$ corresponds to the rotation of the body, $O\,B\,B$ to the relative, and $O\,C\,C$ to the absolute, rotation of the eyes. No one, on repeating the experiment, can avoid the conclusion that we are concerned here with an automatic (unconscious) movement of the eyes, reflexively excited by the rotation of the body. This movement disappears as soon as the passive rotation is no longer felt. How it is brought about remains, naturally, to be investigated. A simple way of looking at it would be that there are two antagonistic organs of innervation, and that the stimulus which reaches them both uniformly when the body rotates, is answered by one with a uniform stream of innervation, while the other delivers its impulse of innervation only after the lapse of a certain time, like a rain-gauge suddenly tipping up when it is full. For us it suffices, provisionally, to know that this automatic, unconscious compensational movement of the eye is actually present.

The compensational wheel-like movement of the eyes, which takes place when the head is inclined to one side, is well-known. Nagel [15] has proved that it amounts to from one-tenth to one-sixth of the angle of inclination of the head. And recently Breuer and Kreidl [16] have made similar experiments in the rotatory apparatus also, with the following results:

"We have a sensation, as Purkynie and Mach have maintained, of the direction of the acceleration of masses. When this direction is changed by the interference of a horizontal acceleration affecting the body on one side, a wheel-like movement of the eyes is set up and persists as long as the interference lasts. It amounts to 0.5 or 0.6 of the angle of deviation. The rotation of visual space, and the appearance of vertical lines as oblique, which takes place under such conditions, depend therefore on an actual but unconscious rotation of the eyes."

I must also mention here two papers by Crum Browne [17] on compensating movements of the eyes.

6.

The slower unconscious compensating movement of the eye (the jerking movement leaves behind it no optical impression) is thus the reason why, when the head is turned, objects seem to retain their position a fact which is very important for orientation. If, now, in turning our head, we also voluntarily turn the eyes in the same direction, fixing them upon one object after another, we must overcompensate the automatic, involuntary innervation by the voluntary innervation. We need the same innervation as if the whole angle turned through were described by the eye alone. In this way is explained why, when we turn about, the whole optical space appears to us a continuum and not an aggregation of fields of vision; and why, at the same time, the optical objects remain stationary. That which we see of our own body, in turning, we see, for obvious reasons, optically in motion.

Thus we arrive at the practically valuable conception of our body as in motion in a fixed space. We understand why it is that, in our numerous turnings and ramblings in the streets and in buildings, and in our passive turnings in a wagon or in the cabin of a ship, even in the dark, we do not lose our sense of direction, though it is true that the primary co-ordinates from which we started gradually sink unnoticed into unconsciousness, and we soon begin to reckon from new objects around us. That peculiar state of confusion as to locality in which we sometimes find ourselves on suddenly awaking at night, where we look about helplessly for the window or the table, is probably due to dreams of movement immediately preceding our awaking.

Similar phenomena to those which manifest themselves on the rotation of the body make their appearance in connexion with the movements of the body generally. If I move my head or my whole body sidewise, I do not lose sight of an object on which my eyes rest. The latter seems to continue motionless, while the more distant objects undergo a displacement in the same direction as that of the body, and nearer objects a parallactic displacement in the opposite direction. The parallactic displacements to which we are accustomed are perceived, but do not cause us any disturbance and are correctly interpreted. But in the monocular inversion of a Plateau wire -net, the parallactic displacements, which in the present case are unusual as regards amount and direction, immediately attract the eye, and apparently present to us a revolving object. [18]

7.

When I turn my head, I not only see that part of it turning which I am able to see (as will be immediately understood from the foregoing) but I also feel it turning. This is due to the fact that conditions exist in the province of touch which are entirely analogous to those in the province of sight. [19] When I reach out my hand to grasp an object, a sensation of touch is combined with an innervation. If I look towards the object, a luminous sensation is substituted for the sensation of touch. Even where objects are not touched, skin-sensations may always be perceived when the attention is turned to them, and these, combined with changing innervations, also yield the conception of our body as in motion, which quite accords with that acquired by optical means.

Thus, in active movements, the skin-sensations are delocalized, as we may briefly express it. In passive movements of the body, reflex, unconscious compensatory innervations and movements of compensation make their appearance. In turning round to the right, for example, my skin-sensations are compounded with the same innervations as would be combined with the touching of objects in turning to the right. I feel myself turning to the right. If I am passively turned toward the right, the reflex endeavour arises to compensate the turning. I either actually remain stationary and feel myself at rest, or I repress the motion toward the left. But for this I need to exert the same voluntary innervation as for an active turning to the right, which has also the same sensation as its result.

8.

At the time when my work on the *Sensations of Movement* was written, I had not yet attained to a thoroughly comprehensive view of the simple relation here described. I encountered, consequently, difficulties in the explanation of certain phenomena, observed partly by Breuer and partly by myself, which are now easy of explanation, and which I will briefly notice. If an observer be shut up in a closed receptacle, and the receptacle be set in rotation toward the right, it will appear to the observer as optically in rotation, although every ground of inference for relative rotation is wanting. If his eyes perform involuntary compensatory movements to the left, the images on the retina will be displaced, with the result that he has the sensation of movement toward the right. If, however, he fixes his eyes upon the receptacle, he must voluntarily compensate the involuntary movements, and thus again he is conscious of movement towards the right. It is plain, therefore, that Breuer's explanation of the apparent motion of optical vertigo is correct, and also that this movement cannot be made to disappear by the voluntary fixation of the eyes. The remaining cases of optical vertigo noticed in my work may be disposed of in like manner. [20]

In voluntary forward motion or rotation, we have not only a sensation of every single successive position of the parts of our body, but also the much more simple sensation of movement forward or of turning round. As a fact, we do not compound the notion of forward movement from the notions of the various individual movements of the legs, or at least we do not need to do so. There are cases, indeed, in which the sensation of forward movement is undoubtedly present while that of the movements of the legs is equally undoubtedly lacking. This is true, for instance, of a railway journey, or even of the thought of a journey, and may occur also in recalling a distant place, etc. The only possible explanation of this can be that the will to move forward or to turn about, which furnishes to the extremities their motor impulses, impulses which may be further modified by particular innervations, is of a comparatively simple nature. The conditions existing here are probably similar to, although more complicated than, those connected with the movements of the eyes, which Hering has so felicitously interpreted, and to which we shall presently return.

We shall scarcely go far wrong if we suppose that the comparatively simple motor-sensations [21] stimulated from the labyrinth of the brain stand in the closest connexion with the will to move. These motor-sensations may also correspond to the feelings of direction which Riehl has postulated and investigated. [22] The blind man has them equally with the man of normal sight, and they probably form an important part of the foundation on which the understanding of tactual space reposes.

I summed up a series of observations on optical sensations and motor-sensations in these words: "It looks as if visual space would turn into a second space, which is held to be immovably stationary, although this second space is characterized by complete absence of visibility." The space which is built up from motor-sensations seems, in fact, to be the original space. [23]

At that time I was preoccupied with physical methods of thought, and was consequently inclined to believe that sensations of progressive acceleration behaved in a manner completely analogous to sensations of angular acceleration. And in fact every physicist who studies this subject is liable to think at once of the three equations for the rotatory movement of a body and the three equations for its movement of translation. I believed further, that, in accordance with the principle of specific energy, we ought to assume special sensations of the position of the head.

Breuer, [24] in a later piece of research, has made it probable that the sensations of progressive acceleration vanish very much more quickly than those of angular acceleration, and that perhaps the organ of the former, at any rate in human beings, is atrophied. Further, Breuer finds that, except for the semicircular canals, B, the otolithic apparatus, O, with its planes of sliding corresponding to the planes of the semicircular canals, is the only organ adapted to the signalizing of progressive accelerations and position simultaneously. The three components of gravity corresponding to the three planes of sliding characterize the position of the head. Every alteration of the position alters these components, and instantaneously sets the apparatus of the semicircular canals going. But progressive accelerations alter these components without making any demands on the semicircular canals. Consequently, according to Breuer, the three combinations, O alone, $O + B$, and B alone, would suffice for the decision of all cases. Thus this theory, if it can be maintained, would be an important simplification.

If I still had it in my power to carry out experiments, I would submit the motor-sensations in themselves to a renewed and thorough investigation. The difference in the behaviour of the sensations of angular and progressive accelerations now seems to me significant. Acceleration of rotation gives rise to a sensation which, long after the acceleration has become nil, persists with a decreasing force which can be quantitatively [25] measured. Progressive acceleration is felt in its pure form only in the case of vertically accelerated falling or rising. When the acceleration vanishes the sensation also disappears quickly. The simplest means of producing a constant acceleration in a constant direction with respect to the body is by uniform rotation. We soon cease to have any sensation of uniform rotation. But a constant centrifugal acceleration also does not evoke the illusion of flying away in its direction, but rather calls up the sensation of changed position, which sensation again vanishes immediately with the disappearance of the centrifugal acceleration. Does this mean that progressive acceleration exhausts itself as a stimulus, or that when the stimulus becomes constant the sensation changes in character? In that case we shall have to suppose that the sensation is composed of two elements.

We have sensations, not of uniform motion, but only of acceleration. To the elements of the change in progressive and angular velocity there correspond elements of the motor-sensations; and, of these, those sensations at any rate which correspond to angular velocities persist with gradually decreasing force, and moreover are algebraically summable just like the sensations corresponding to progressive velocities; so that a sensation p, corresponding to the total change in velocity, and consequently to the velocity attained v, is correlated with a movement, usually of velocity nil upwards, set up in a short time. [26] Now the aggregate of sight and-touch-impressions that we passed in review increases with p and with the time t. We need not therefore be surprised that experience teaches us to interpret p conceptually as a velocity and pt as a path, although of course p in itself has nothing to do with concepts of spatial measurement. In this way it seems to me that we get rid of the last remaining paradox in the theory of the motor sensations. This paradox still troubled me in 1875, and I see that it has also troubled others. [27]

9.

The following experiments and reflexions, which form a sequel to an earlier publication of mine, [28] will perhaps assist us in obtaining a correct view
of these phenomena.

If we take our stand upon a bridge, and look fixedly at the water flowing beneath, we shall generally have the sensation of being ourselves at rest, whilst the water will seem in motion. Prolonged gazing, however, as is well known, almost invariably results in

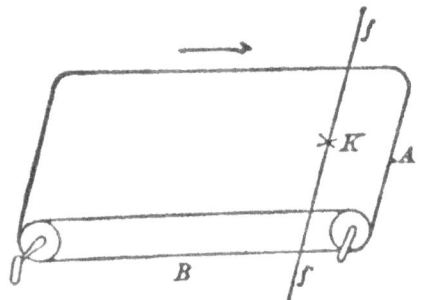

Fig. 18.

84

the sensation that suddenly the bridge, with the observer and his whole environment, begins to move in the direction opposite to that of the water, while the water assumes the appearance of being at rest. [29] The relative motion of the objects is in both cases the same, and there must therefore be some adequate physiological reason why at one time one, and at another another, part of them is felt to move. In order to investigate the matter at my leisure, I had the simple apparatus constructed which is represented in Fig. 18. An oil-cloth of simple pattern is drawn horizontally over two rollers, two metres long and fixed three metres apart in bearings, and is kept in uniform motion by means of a crank. Across the oil-cloth and about thirty centimetres above it, is stretched a string *ff*, with a knot *K*, which serves as a fixation-point for the eye of the observer stationed at *A*. Now, if the oil-cloth be set in motion in the direction of the arrow, and the observer follow the pattern with his eyes, he will see it in motion, himself and his surroundings at rest. On the other hand, if he gazes at the knot, he and the whole room will presently appear in motion in the contrary direction to the arrow, while the oil-cloth will stand still. This change in the aspect of the motion takes more or less time according to the mental condition of the observer, but usually requires only a few seconds. If we once get the knack of it, the two impressions may be made to alternate with some rapidity and at will. Every following of the oil-cloth brings the observer to rest, every fixation of *K*, or non-attention to the oil-cloth, by which its pattern becomes blurred, sets the observer in motion. Two investigators, for whom I have the deepest respect, do not agree with me as to the result of this experiment in the circumstances indicated. One is William James, [30] the other Crum Brown. [31] I have performed the experiment over and over again with the same result. I am at present not in a position to carry out experiments, and must consequently renounce any idea of a new test; but the method of afterimages described by Brown seems to have much to recommend it. I pass over for the moment the different possible theoretical explanations of the experiment.

10.

This phenomenon, of course, must not be confounded with the familiar Plateau-Oppel phenomenon, which is a *local* retinal effect. In the preceding experiment, the entire environment, so far as it is distinctly visible, is in motion, whilst in the latter a moving veil is drawn along in front of the object, which is at rest. The attendant stereoscopic phenomena, - for example, the appearance of the thread and knot underneath the transparent oil-cloth, - are quite immaterial in this connexion.

In my book on *Bewegungsempfindungen* (p. 63) I made it clear that the Plateau-Oppel phenomenon was the result of a peculiar process, which had nothing to do with the other sensations of movement. I wrote there as follows:

"We must therefore suppose that, during the movement of an image on the retina, a *peculiar* process is excited which is absent during rest, and that in the case of movements in opposite directions, very similar processes are excited in similar organs, processes which are, however, mutually exclusive, so that with the commencement of the one, the other must cease, and with the exhaustion of the one, the other begins."

This statement of mine seems to have been overlooked by S. Exner and Vierordt, who subsequently expressed similar views on the same subject.

11.

Before we proceed to the explanation of the experiment (Fig. 18), it will be well to introduce a few variations. An observer stationed at B seems, under the same conditions, to be speeding, with all his surroundings, towards the left. We now place above the oil-cloth $T\,T$, Fig. 19, a mirror $S\,S$, inclined at an angle of 45° to the horizon. We observe the reflexion $T'\,T'$ in $S\,S$, after having placed on our nose a shade $n\,n$, which intercepts the direct view of $T\,T$ from the eye, O. If $T\,T$ moves in the direction of the arrow, while we are looking at K', the reflexion of K, we shall presently fancy ourselves sinking downward with the whole room, whereas if the motion be reversed, we shall seem to ascend as if in a balloon. [32] Finally, the experiments with the paper drum, which I have elsewhere described, [33] and to which the following explanation also applies, should be cited here. None of these phenomena are purely optical, but all are accompanied by unmistakable motor sensations of the whole body.

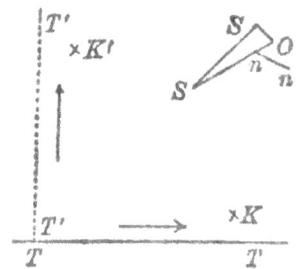

Fig. 19.

12.

What form, now, must our thoughts take on, in order to acquire the simplest explanatory setting for the preceding phenomena? Objects in motion exert, as is well known, a peculiar motor stimulus upon the eye, and draw our attention and our gaze after them. If the eye really follows them, we must assume, from what has gone before, that the objects appear to move. But if the eye is kept for some time at rest in spite of the moving objects, the constant motor stimulus proceeding from the latter must be compensated by an equally constant stream of innervation flowing to the motor apparatus of the eye, exactly as if the motionless point on which the eyes rest were moving uniformly in the opposite direction, and we were following it with our eyes. But when this process begins, all motionless objects on which the eyes are fastened must appear in motion. It is obviously unnecessary that this stream of innervation should always be consciously and deliberately called into action. All that is requisite is that it should proceed from the same centre and by the same paths as voluntary fixation.

No special apparatus is necessary for observing the foregoing phenomena. They are to be met with on all hands. I walk forward by a simple act of the will. My legs swing to and fro without my having to attend to them particularly. My eyes are fixed steadfastly upon my goal without suffering themselves to be drawn aside by the motion of the retinal images consequent upon progression. All this is brought about by a single act of the will, and this act of the will itself is the sensation of forward movement. The same process, or at least a part of it, must also be set up, if the eyes are to resist for any length of time the stimulus of a mass of moving objects. Hence the motor sensations experienced in the above experiments.

The eyes of a child in a railway train will be observed to follow almost uninterruptedly and with a jerking motion the objects outside, which appear to it to be running. The adult has the same sensation if he will passively yield himself to the natural impressions. If I am riding forwards, the whole space to my left, for obvious reasons, rotates, in the direction of the hands of a watch, about a very distant vertical axis and the space to my right does the same, but in the opposite direction. Only when I resist following the objects with my eyes, does the sensation of forward motion arise.

13.

My views regarding the sensations of movement have been repeatedly attacked, as is well known, but invariably the adverse arguments have been aimed solely at the *hypothesis,* to which I attached comparatively little importance. That I am ready and willing to modify my views in accordance with newly discovered facts, the present work will testify. The decision as to how far I am in the right I will cheerfully leave to the future. On the other hand, I should like to mention that observations have been made that strongly favor the theory propounded by myself, Breuer, and Brown. To these belong, first, the facts collected by Dr. Guye of Amsterdam (*Du Vertige de Ménière: Rapport lu dans la section d'otologie dit congrès periodique international de sciences medicales à Amsterdam,* 1879). Guye observed, in diseases of the middle ear, that reflex turnings of the head were induced when air was blown into the cavity of the tympanum, and found a patient who was able to state exactly the direction and number of the turnings which he had felt during the injection of liquids. Prof. Crum Brown ("On a Case of Dyspeptic Vertigo," *Proceedings of the Royal Society of Edinburgh,* 1881-1882), has described an interesting case of pathological vertigo observed in himself, which admitted of explanation, as a whole, by the increased intensity and lengthened duration of the sensation incident upon every turning of the body. But most remarkable of all are the observations of William James ("The Sense of Dizziness in Deaf-Mutes," *American Journal of Otology,* Vol. IV., 1882). James discovered in deaf-mutes a striking and relatively general insensibility to the dizziness of whirling, often great uncertainty in their walk when their eyes were closed, and in many cases an astonishing loss of the sense of direction on being plunged under water, in which case there always resulted alarm and complete uncertainty as to up and down. These facts speak very strongly in favor of the view, which naturally follows from my conception, that in deaf-mutes the sense of equilibrium proper is considerably degenerated, and that the two other senses that give direction, the sense of sight and the muscular sense (the latter of which loses all its points of reference when the weight of the body is neutralized by immersion in water), are rendered proportionately more necessary.

It is impossible to maintain the view that we arrive at knowledge of equilibrium and of movement solely by means of the semi-circular canals. On the contrary, it is extremely probable that lower animals, in whom this organ is entirely wanting, also have sensations of movement. I have not yet been able to undertake experiments in this direction. But the experiments which Lubbock has described in his work, *Ants, Bees, and Wasps,* become much more comprehensible to me on the assumption of sensations of movement. As experiments of this sort may be interesting to others, it will not be amiss perhaps to consider an apparatus which I have briefly described before (*Anzeiger der Wiener Akademie,* 3oth December 1876). Other apparatuses of the same kind have since been constructed by Govi and Ewald. They have been called "cyclostats."

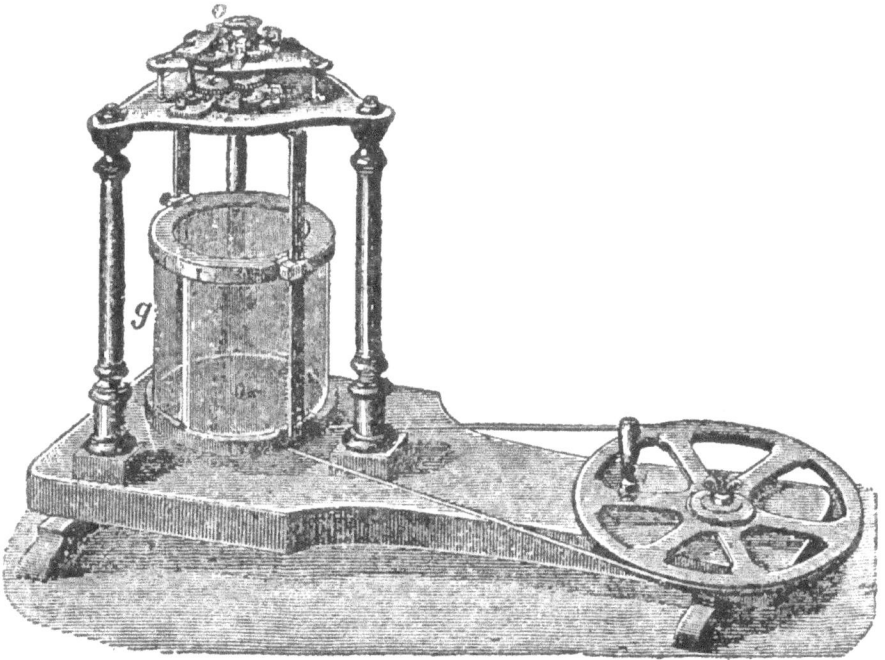

Fig. 20.

The apparatus serves for the observation of the conduct of animals while in rapid rotation. Since, however, the view of the animal will necessarily be effaced by the rotatory motion, the passive rotation must be optically nullified and eliminated, so that the active movements of the animal alone shall be left and rendered observable. The optical neutralization of the rotatory motion is attained simply by causing a totally reflecting prism to revolve, with the aid of gearing, above the disk of the whirling machine, about exactly the same axis, in the same direction, and with half the angular velocity of the disk.

Fig. 20 gives a view of the apparatus. On the disk of the whirling machine is a glass receiver, g, in which the animals to be observed are enclosed. By means of gearing the eyepiece o is made to revolve with half the angular velocity and in the same direction as g. The following figure gives the gearing in a separate diagram. The eyepiece oo, and the receiver gg, revolve about the axis AA, while a pair of cog-wheels, rigidly connected together, revolve about BB. Let the radius of the cog-wheel act, rigidly connected with gg, be = r. Then r is the radius of M, and $2r/3$ is the radius of cc t but the radius of dd is = $4r/3$, wherewith the desired relation of velocity between oo and gg is obtained.

In order to centre the apparatus, a mirror S, provided with levelling-screws, is laid upon the bottom of the receiver and so adjusted that, on rotation, the reflexions in it remain at rest. It is then perpendicular to the axis of rotation. A second small mirror, S', in the silvering of which is a small hole L, is so adjusted to the open tube of the eyepiece, with its reflecting surface downward, that, on rotation, the images seen through the hole, in the mirrored reflexion of S' in S, remain motionless. Then S' stands perpendicular to the

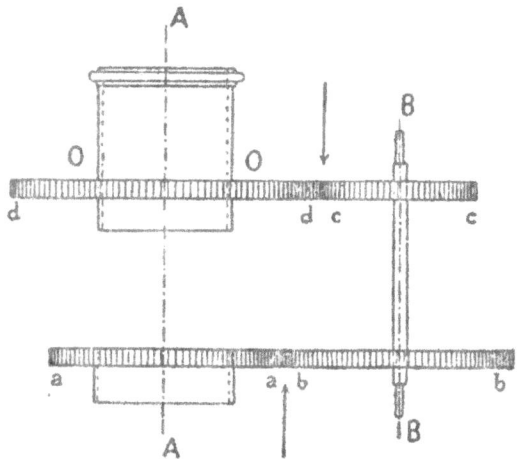

Fig. 21.

axis of the eyepiece. With the aid of a brush we may now mark upon the mirror S a point P, whose position is not altered on rotation (a result which is easily accomplished after a few trials), and so place the hole in the mirror S' that it also remains stationary on rotation. In this way J points on both axes of rotation are found. If now by means of screws we so adjust the eyepiece, that, on looking through the hole in S', the point P on S and the reflexion of L in S' (or really the many reflexions of P and L) fall on the same spot, then the two axes are not only parallel but coincident.

The simplest eyepiece that can be employed is a mirror whose plane coincides with the axis, and I adopted this device in the initial form of my apparatus. But one-half of the field of vision is lost by this method. A prism of total reflexion, therefore, is much more advantageous. Let ABC (Fig. 22) represent a plane section of such a prismatic eyepiece cut perpendicularly to the planes of the hypothenuse and the two sides. Let this section include, also, the axis of rotation $ONPQ$, which is parallel to AB. The ray which passes along the axis QP must, after refraction and reflexion in the prism, proceed again along the axis NO and will meet the eye O in the prolongation of the axis. This done, the points of the axis can suffer no displacement from rotation, and the apparatus is centred. The ray in question must accordingly fall at M, the middle point of AB, and, hence, since it falls on crown glass at an angle of incidence of 45°, will meet AB at about 16° 40'. Therefore, OP must be distant about 0.115 AB from the axis, a relation which can best be obtained by trial, by so moving the prism in the eyepiece that oscillations of the objects in gg during rotation are eliminated.

Fig. 22 also shows the field of vision for the eye at O. The ray OA, which falls vertically upon AC, is reflected at AB in the direction AC and passes out towards S. The ray OR, on the other hand, is reflected at B and emerges, after refraction, in the direction of T.

The apparatus has hitherto proved quite sufficient for my experiments. If a printed page is placed in gg, and the apparatus turned so rapidly that the image on the retina is entirely obliterated, one can easily read the print through the eyepiece. The inversion of the image by reflexion could be removed by placing a second, stationary reflecting prism above the revolving

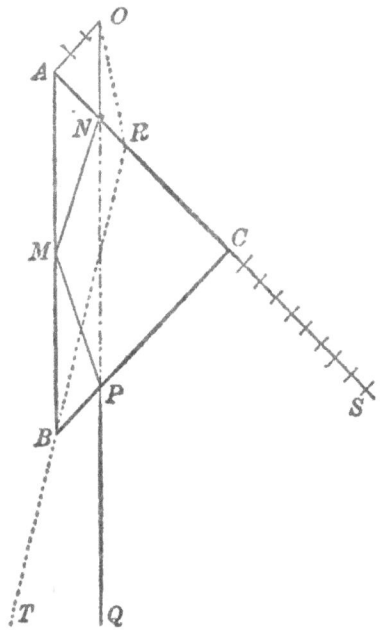

Fig. 22.

prism of the eyepiece. But this complication appeared to me unnecessary.

With the exception of a few physical experiments, I have hitherto undertaken rotation-experiments only with various small vertebrates (birds, fishes), and have found the data given in my work on *Motor Sensations* fully confirmed. However, it would probably be of advantage to make similar experiments with insects and other lower animals, especially with marine animals.

Such experiments have subsequently been carried out, with most instructive results, by Schäfer (*Naturwissenschaftliche Wochenschrift*, No. 25, 1891), by Loeb (*Heliotropismus der Tiere*, Würzburg, 1890, p. 117), and by others. In my lecture "On Sensations of Direction" (*Schriften des Vereins zur Verbrei-*

tung naturwissenschaftlicher Kenntnisse in Wien, 1897, and *Populärwissen-schaftliche Vorlesungen, 3rd* edition, 1903) will be found the remainder of what I have to say on the sense of direction. But I should like to refer particularly to Breuer's researches on the otolithic apparatus, to Pollak and Kreidl's experiments on deaf-mutes, and above all to Ewald's work of fundamental importance, *Ueber das Endorgan des Nervus octavus* (Wiesbaden, 1892). In the third volume of the *Handbuch der Physiologie des Menschen* (1905), by W. Nagel, there is a full account of the "theory of the sensations of position, movement, and resistance." Since for some years past I have not been in a position to follow the experimental work that has been done in this department with any closeness, I have asked Professor Josef Pollak to give an account here of so much of the most recent work as is likely to interest readers of this book. Dr. Pollak has very kindly complied with my request, and the following paragraphs 14-19 are from his pen.

14.

The results in the course of the last ten years of morphological research and of research in comparative and experimental physiology in connexion with the labyrinth of the ear (the cochlea, the semicircular canals, and the otolithic apparatus) have been almost without exception favorable to the Mach-Breuer hypothesis.

It may now be taken as proved that the organ of hearing is constituted by the cochlea alone, and that the vestibular apparatus has no acoustic functions. A complete proof of this has been furnished by Biehl, [34] who, by intracranial operations on sheep, succeeded in severing the vestibular branch of the acoustic nerve without injuring the ramus cochlearis; the result was that disturbances of equilibrium were produced, though the sense of hearing remained unaffected. Further, that part of the theory of the static function of the labyrinth is firmly founded and scarcely open to attack, which regards the semicircular canals as sense-organs that serve the perception of turnings of the head (and, mediately, of the body), especially since this hypothesis has received at the hands of Breuer important modifications based on his anatomical studies of the epithelial hairs of the ampullae. [35]

This hypothesis now runs as follows:

"Persistent uniform rotations are not felt, however rapid they are; but the beginning and end of the rotation, acceleration and retardation, are felt. The ampullary apparatus is not affected by angular velocities that persist, but only by positive and negative angular accelerations. These cause a momentary displacement of the endolymphring and of the cupula terminalis (which as a consistent mass holds the epithelial hairs together in a constant figure of fixed shape), and, concomitantly, set up a tension of the cell-hairs and an excitement of the terminal apparatus of the nerves on one side of the crista involved. As long as these excitations last, they give rise to a sensation of rotation, which persists until the contrary impulse of negative acceleration, when the rotation stops, or the gradual effect of the elasticity of the stretched complexes, restores the normal state of things again."

The system of semicircular canals, moreover, possesses, like all other sense-organs, the property of giving rise not only to sensations but also to reflexes (Breuer, Delage, Nagel). Prominent among reactive organs are the muscles of the eyes, which communicate rotations to the eyes when the body rotates.

15.

Previously, however, the opinion had been conjecturally put forward by Mach that progressively accelerated motion could exercise no influence on the lymph enclosed in the semicircular canals; he had also suggested that special organs exist in the labyrinth for the perception of accelerations and for the sensation of the position of the head. Breuer then succeeded in making it at least very probable that this function belongs to the otolithic apparatus. He supposes that the otoliths exert, by means of their weight, a definite pressure on the hair-cells underneath them. Every inclination of the head must change the position of the sacculus, and consequently that of the senseepithelia also. By determining the position of the directions in which the otoliths slide for different positions of the head, Breuer showed that an unambiguous pronouncement as to the position of the head is only made possible by the co-operation of the two sacks. "For every position of the head there is only one definite combination of magnitudes of gravitation of the otoliths in the four maculae. When, as we suppose, the gravitation of the otoliths is felt, then every position of the head is characterized by a definite combination of these sensations." In the case of acceleration in a straight line, every shock that causes motion will evoke, owing to the inertia of the otolithic masses, a relative acceleration of these masses in the opposite direction, this relative acceleration representing the adequate sensational stimulus.

Heuristically, this part of the hypothesis is now on a very firm footing. It has become the basis of research on the lower animals in which otoliths alone occur, and in the case of higher animals also it has pointed the way to the isolation and experimental testing of the functions of the otoliths.

From the mass of facts concerning the lower animals, which have been discovered of recent years, I will only select a few pregnant instances. The phenomena that result on the removal of the otoliths have been studied, also the behaviour of animals under rotation and the compensatory movements. The experiments of Prentiss, [36] are particularly interesting. He first repeated Kreidl's famous experiments in compelling sloughing Crustacea to absorb "iron" otoliths; he confirmed the fact that the behaviour of these towards magnets is in accordance with the theory. But he was also so fortunate as to obtain observations on free-swimming larvse of lobsters, which had been deprived of the power of growing otoliths after they had sloughed their skin. He was able to convince himself that they present the same phenomena as full-grown shrimps, from which the otoliths have been removed: they roll from one side to another, swim with belly upwards, are more easily turned over on to their backs than normal larvae, and, when they are blinded, the loss of equilibrium is still more striking. The same writer also describes as follows the behaviour of a crustacean (*Virbius zostericula*) in which the statocyst is normally absent.

"It is not a free-swimming form, but attaches itself to grasses, in the positions that are independent of gravity. If it is compelled to swim, it does so in a very uncertain manner, but generally back upwards. It is easily turned over on to its back, and, once in this position, is very slow at setting itself right. Its uncertain manner of swimming is reminiscent of that of other Crustacea after their statocysts have been destroyed. If the eyes are covered with lampblack, all sense of direction in swimming is lost."

The experiments of Prentiss recall those of K. L. Schafer. [37] Schafer rotated the larvae of frogs, and discovered that the first appearance of rotatory vertigo coincides in time with the completion of the formation of the semicircular canals.

Ach's [38] researches on frogs are important. He discovered that the otoliths are connected with the lid-reflex of the crossed side; from the fact that, in the case of a frog deprived of its otoliths and subjected to rapid movement, the lid-reflex vanishes both horizontally and vertically, he concluded that the function of the otoliths is to subserve displacements of the body in a straight line in space.

16.

On the other hand, the wheel-like movements of the eyes that take place when the position of the head undergoes a series of changes, and the nystagmic movements caused by rotation and by passing a galvanic current through the head, have long been known and have been sufficiently analysed. The typical movements of the head, and the jerking movements of the eyes which are repeated at regular intervals when the head is continuously rotated or when a galvanic current is passed through it, and which can also be easily felt through the closed eyelids, are sure objective signs of vertigo. Nystagmus of eyes and head is completely absent in animals without a labyrinth, as has been shown by Ewald in the case of doves, and by Breuer in the case of cats, whose nervous octavus had been severed on both sides. Breuer and Kreidl have proved that the optical distortion of the vertical, experienced by anyone who rides in a whirling chair or sits in a railway train passing quickly enough over a sharp curve, depends upon a real wheel-like movement of the eyes. Again, we owe to Breuer the proof that individual ampullae, even when isolated, can be galvanically stimulated; when that is done, they produce a movement of the head in the plane of the canal involved, whereas, according to Breuer, the consequence of *diffused* stimulation is the so-called galvanotropic reaction, consisting in an inclination of the head towards the anode.

So much having been premised, the phenomena observed by James, [39] Kreidl [40] and Pollak [41] as resulting with deaf-mutes when affected with rotatory or galvanic vertigo, can easily be explained on the Mach-Breuer theory. According to Mygind, [42] out of 118 deaf-mutes subjected to an anatomical examination, pathological changes of the vestibulary apparatus were present in 56 per cent.; 50 to 58 per cent, of the deaf-mutes experimented on by Kreidl felt no rotatory vertigo; 21 per cent, of those on whom Kreidl reproduced the conditions of Mach's experiment with the whirling chair, did not succumb to the illusion as to position with respect to the vertical, which is inevitable in normal persons: they also without exception displayed, when rotated, no reflex movements of the eyes. The explanation of the lower percentage is that, according to Mygind's statistics, the semicircular canals are more frequently found to be diseased than the vestibule.

Pollak found that 30% of the deaf-mutes he investigated experienced no galvanic vertigo, and that most of the deaf-mutes who display no eye-movements and no illusion as to the vertical when placed on the rotating platform or the whirling chair, are also devoid of the characteristic symptoms of galvanic vertigo. Further researches by Strehl, Kreidl, Alexander and Hammerschlag confirmed these facts; the three latter discovered, further, that, when the deaf-mutes were divided into congenital deaf-mutes and those with acquired deafness, an extremely high percentage of the former (in Kreidl and Alexander's experiments 84%, in Hammerschlag's 95%) displayed normal galvanic reaction, while only 29% of the subjects with acquired deafness reacted normally to the galvanic current.

Congenital deaf-mutes, *i.e.*, those with hereditary degeneration, behave in this connexion in the same way as Japanese dancing mice, the explanation of whose physiological behaviour lies, as Kreidl and Alexander [43] have shown, in their anatomical structure.

These mice are completely deaf, and progress in a sprawling, hobbling fashion; to a superficial observer their power of equilibrium seems unimpaired; but if one tries experimentally to get them to move along a narrow path, the high degree in which their power of balance is defective becomes immediately obvious. They are free from rotatory vertigo, but when a galvanic current is passed through their heads they behave like normal animals. Anatomical examination gives the following results. Destruction of the papilla basilaris cochleae, advanced emaciation of the ramus inferior of the eighth nerve, advanced atrophy of the spiral ganglion, destruction of the macula sacculi, medium emaciation of the branches and roots of the ramus superior and medius of the eighth nerve, and medium diminution of both vestibular ganglia.

17.

Among recent experiments in the field of comparative physiology. those of Dreyfuss [44] seem to me very remarkable. He observed the behaviour of normal guinea-pigs, and of guinea-pigs deprived of their labyrinths (operated on one side only and on both sides), when placed on a rotating platform, his special object being to study the compensatory movements of the eyeball and the head. He records a striking difference in the behaviour under rotation of the operated animal as contrasted with the behaviour of the intact animal. The animal that has been deprived of both labyrinths remains motionless in one place under rotation; it displays no displacement of the longitudinal axis of the vertebral column, and no nystagmus of head or eyes. It is unconscious of the rotation. To prove this, Dreyfuss devised the following experiment in feeding the guinea-pigs. If four of them, - one normal, one with the left, one with the right, and one with both labyrinths destroyed, - are placed on the rotating platform, and the experimenter waits until they have all begun to feed, the normal guinea-pig stops eating during rotation; the guinea-pig without the right labyrinth goes on eating while the rotation is to the right, and stops when it is to the left; the one without the left labyrinth goes on eating when the rotation is to the left, and stops when it is to the right; and the one with neither labyrinth goes on eating whichever the direction of rotation. Breuer and Kreidl obtained analogous results from comparative experiments with normal and acoustically defective cats.

18.

Morphologically, and from the teleological standpoint, Alexander's work on the organs of equilibrium and hearing in animals with congenitally defective visual apparatus, the mole (*Talpa europaea*) and the blind mouse (*Spalax typhlus*), is interesting. [45]

It is well-known that, in comparison with the lower animals, the vestibular apparatus in the higher animals and in man is defectively developed. In the case of all animals that are able to move in air or water, we find three nerveends carrying statoliths; in the higher mammals there are only two. As regards the higher animals, Mach and Breuer have repeatedly emphasized the fact that they are far from meaning to imply that the labyrinth alone furnishes the sensations necessary for the maintenance of equilibrium; rather it co-operates to this end with the sense of pressure and the muscular sense, as well as with the sense of sight." It has never been denied, and is indeed quite certain, that absence or loss of the labyrinth-sensations can to a large extent be replaced by the other sense-perceptions just mentioned, so that, as Ewald pre-eminently has shown, the major functions of the maintenance of equilibrium, such as walking and standing, can be adequately performed even when the labyrinth-function has been lost, or when there is some congenital defect in it. We see this, not only in the_ case of operated animals, but also in that of those deaf mutes in whom we have reason to assume some lesion of the semicircular canals (Breuer). However, James and Krei'dl have shewn that deafmutes who are not subject to rotatory vertigo are very unskilful in all the more delicate problems of balance.

The mole, on the other hand, is an animal whose movements take place principally underground, though the surface on which it moves is solid; moreover it dispenses almost completely with any sense of orientation by means of the organ of sight, and Alexander has shown that this is fully compensated by its exquisite power of balance. This is anatomically expressed by the unusual size of the terminal nerve cells, by the relatively large number of the sense-cells, and especially by the presence, in the sinus utricularis inferior, of a macula neglecta which is wanting in other mammals, and which, apart from birds and reptiles, has only been found in one other lower mammal, *Echidna aculeata*.

It is Alexander's merit to have proved that, in respect of the structure of its static terminal nerve-cells, *Echidna* represents the hitherto unknown transition from mammals to birds. *Echidna* possesses a cortical organ which, in histological structure, corresponds with that of mammals, whereas in the number of the other terminal points of nerves it agrees with the labyrinth of birds; in addition to the three macular nerve-terminations (macula utriculi, macula sacculi, and macula lagenae) it exhibits a macula neglecta Retzii.

19.

The results of these researches, of which only a small selection has been mentioned, may be summed up as follows. The compensation in the visual field of every movement of the head by means of movements of the eyes, which are carried out also by the blind and by normal persons with their eyes shut; the absence of these movements in many deaf-mutes; the nystagmus of the eyes that takes place under continued rotation; the wheel-like movement of the eyes when the direction of the acceleration of masses in the body is altered by a centrifugal force; rotatory vertigo and its law; the absence "of this vertigo in many deaf-mutes; finally, the identical character of galvanic vertigo in man and in animals, - all these facts serve strongly to confirm the theory of Mach and Breuer, although it cannot be denied that many questions still remain unsolved. As against other hypotheses, such as those of Ewald and Cyon, this theory has at any rate the advantage that, in the case of the ampullary and otolithic apparatus, it provides a clearer explanation of the specific disposition to the adequate stimulus than we have for any other sense organ, and also that, in accordance with it, the two sense-organs in the labyrinth are readily brought under the principle of specific sense-energies (Nagel). In any case the sensation of motion is proved to be a special and peculiar department of sensation.

20.

Professor Pollak's communication here comes to an end.

Without doing violence to the facts described in my book on *The Sensations of Movement,* the preceding observations suggest the possibility of modifying the theoretical view there taken of the facts, as we shall point out in what follows. It remains extremely probable that an organ exists in the head it may be called the terminal organ (*TO*) which reacts upon accelerations, and by means of which we are made aware of movements. To me personally the existence of sensations of movement, of the same nature as other sensations, does not seem doubtful, and I can scarcely understand how anyone, who has really repeated on himself the experiments in question, can deny the existence of these sensations.

But instead of imagining that the terminal organ excites special motor-sensations, which proceed from this apparatus as from a sense-organ, we might also assume that this organ simply disengages innervations after the manner of reflexes. Innervations may be voluntary and conscious or involuntary and unconscious. The two different organs from which these proceed may be designated by the letters *WI* and *UI*. Both sorts of innervation may pass to the oculomotor apparatus (*OM*) and to the locomotor apparatus (*LM*).

Let us now consider the accompanying diagram. We induce by the will, that is by a stimulus from *WI*, an active movement, which passes in the direction of the unfeathered arrows, to *OM* and *LM*. The appropriate innervation, whether it precedes or follows the movement, is directly felt. In this case, therefore, a specific sensation of movement, differing from the innervation, is unnecessary. If the motion in the direction of the unfeathered arrows is a passive one (taking us by surprise), then, as experience shows, reflexes pass from *TO* to *UI*, which produce compensatory movements, indicated by the feathered arrows. If WI takes no part in the process, and the compensation is effected, both the motion and the necessity for motor sensation are absent. But if the compensatory movement is intentionally suppressed, that is, by intervention Fig. 23. from *WI*, then the same innervation is necessary for achieving this result as for active movement, and it consequently produces the same motor sensation.

The terminal organ *TO* is accordingly so adjusted to *WI* and *UI* that upon a given motor stimulus in the first, contrary innervations are set up in the last two. But further, we have to notice the following difference in the relation of *TO* to *WI* and *UI*. For *TO*, the motor excitation is naturally the same whether the movement induced is passive or active. In active movements, too, the innervations proceeding from *WI* would eventually be neutralized by *TO* and *UI*, did not an inhibitory innervation proceed simultaneously with the willed innervation from *WI* to *TO* or *UI*. The influence of *TO* upon *WI* must be conceived as much weaker than its influence upon *UI*. If we should picture to ourselves three animals, *WI*, *UI*, and *TO*, between whom there was a division of labor, such that the first executed only movements of attack, the second only those of defence or flight, while the third filled the post of sentinel, all of whom were united into a single new organism in which *WI* held the dominant position, we should have a conception approximately corresponding to the relation represented. There is much in favor of such a conception of the higher animals. [46]

I do not offer the preceding view as a complete and perfectly apposite picture of the facts. On the contrary, I am fully aware of the defects in my treatment. But the attempt to reduce to one and the same quality of sensation, in accordance with the cardinal principle evolved in our investigation, all sensations of space and movement which arise in the province of sight and touch during change of place, and which, even when locomotion is only remembered, or a distant spot only thought of, arise in a shadowy form, will be found to be not without justification. The assumption that this quality of sensation is the will, so far as the will is occupied with position in space and spatial movement, or that it is innervation, does not forestall future investigation and only represents the facts as they are known at the present time. [47]

21.

From the discussions of the previous chapter relative to symmetry and similarity, we may immediately draw the conclusion that to like directions of lines which are seen, the same kind of innervation-sensations, and to lines symmetrical with respect to the median plane very similar sensations of innervation correspond, but that with looking upwards and looking downwards, or with looking at objects afar off and at objects near at hand, very different sensations of innervation are associated, - as we should naturally be led to expect from the symmetrical arrangement of the motor apparatus of the eye. With this single observation we explain at once a long chain of peculiar physiologico-optical phenomena, which have as yet received scarcely any attention. I now come to the point which, at least from the physical point of view, is the most important.

The space of the geometrician is a mental construction of three-dimensional multiplicity, that has grown up on the basis of manual and intellectual operations. Optical space (Hering's "sight-space") bears a somewhat complicated geometrical relationship to the former. The matter may be best expressed in familiar terms by saying that optical space represents geometrical space (Euclid's space) in a sort of relievo-perspective - a fact which can be teleologically explained. In any event, optical space also is a three-dimensional multiplicity. The space of the geometrician exhibits at every point and in all directions the same properties a quality which is by no means characteristic of physiological space. But the influence of physiological space may nevertheless be abundantly observed in geometry. Such is the case, for example, when we distinguish between convex and concave curvatures. The geometrician should really know only the amount of deviation from the mean of the ordinates.

22.

As long as we conceive the (12) muscles of the eye to be separately innervated, we are not in a position to understand the fundamental fact that optical space is presented as a three-dimensional multiplicity. I felt this difficulty for years, and also recognized the direction in which, on the principle of the parallelism of the physical and the psychical, the explanation was to be sought; but owing to my defective experience in this province, the solution itself remained hidden from me. All the better, therefore, am I able to appreciate the service rendered by Hering, who discovered it. To the three optical space co-ordinates, viz., to the sensations of height, breadth and depth, corresponds according to the showing of this investigator (Hering, *Beiträge zur Physiologie,* Leipzig, Engelmann, 1861-1865; *Die Lehre vom binokularen Sehen,* 1868) simply a threefold innervation, which turns the eyes to the right or to the left, raises or lowers them, and causes them to converge, according to the respective needs of the case. This is the point which I regard as the most important and essential. [48] Whether we regard the innervation itself as the space-sensation, or whether we conceive the space-sensation as before or behind the innervation, - a question neither easy nor necessary to decide, - nevertheless Hering's account throws a flood of light on the psychical obscurity of the visual process. The phenomena cited by myself with regard to symmetry and similarity, moreover, accord excellently with this conception. But it is unnecessary, I think, to dwell any further on this. [49]

[1] So far as I know, the matter treated in the preceding chapter has not yet been discussed, except in three small works of my own, and in Soret's book. The considerations of the present chapter, moreover, are, for me, founded upon those of the preceding chapter. I indicate here the methods by which I have myself reached clear ideas as to the sensation of space, without laying the least claim to that which has been accomplished by others in this direction, particularly by the theory of Hering. The extensive literature of this subject is, moreover, too imperfectly known to me for me to give exact references on all points. The point of Hering's theory which I regard as the most important I will especially notice.

[2] J. Müller, *Vergleichende Physiologie des Gesichtsinnes,* 1826; *Handbuch der Physiologie,* Vol., II., 1840.

[3] *L'Ottica di Claudio Tolomeo,* published by G. Govi, Turin, 1885.

[4] Wheatstone, "Contributions to the Theory of Vision," *Philosophical Transactions 1838,* 1852.

[5] Panum, *Untersuchungen über das Schen mit zwei Augen,* 1858.

[6] Hering, *Beiträge zur Physiologie,* 1861-1865; *Archiv für Anatomie und Physiologie,* 1864, 1865; "Der Raumsinn und die Bewegungen des Auges," in Hermann's *Handbuch der Physiologie,* Vol. III., I, 1879.

[7] Among the works of younger investigators connected with Hering's researches, those of F. Hillebrand are of particular interest for psychology.

[8] Stumpf, *Der psychologische Ursprung der Raumvorstellungen,* 1873.

[9] *Vergleichende Physiologie des Gesichtssinnes,* pp. 106, sqq.

[10] Tschermak, "Ueber anomale Sehrichtungsgemeinschaft der Netzhäute bei Schielenden," Graefe's *Archiv,* XLVII., 3, p. 508; Tschermak, *Ueber physiologische und pathologische Anpassung des Auges,* Leipzig, 1900; Schlodtmann, "Studien über anomale Sehrichtungsgemeinschaft bei Schielenden," Graefe's *Archiv,* LI., 2, 1900.

[11] Shortly after finishing my *Grundlinien der Lehre von den Bewegungsempfindungen* (1875).

[12] W. James, *The Principles of Psychology,* II., p. 509.

[13] I retain the expression which first immediately suggested itself to me (1875), with no intention of forestalling future inquiry. Here and in what follows I leave it an open question whether innervation is a consequence of space-sensation, or *vice versa.* They are certainly closely connected.

[14] For a different explanation see Lipps, *Zeitschrift für Psychologie und Physiologie der Sinnesorgane,* Vol. I., p. 60.

[15] Nagel, "Ueber Kompensatorische Raddrehungen der Augen," *Zeitschrift für Psychologie und Physiologie der Sinnesorgane,* Vol. XII., p. 338.

[16] Breuer and Kreidl, "Ueber scheinbare Drehung des Gesichtsfeldes während der Einwirkung einer Zentrifugalkraft," Pflüger's *Archiv,* Vol. LXX., p. 494.

[17] Crum Brown, "Note on Normal Nystagmus," *Proc. of the Royal Society of Edinburgh,* 4th Feb. 1895: "The Relation between the Movements of the Eye and the Movements of the Head," *Robert Boyle Lecture,* 13th May 1895.

[18] Compare my "Beobachtungen liber monoculare Stereoscopic," *Sitzungsberichte der Wiener Akademie,* Vol. LVIII., 1868.

[19] The view that the sense of sight and the sense of touch involve, so to speak, the same space-sense as a common element, was advanced by Locke and contested by Berkeley. Diderot also (*Lettres sur les aveugles*) is of opinion that the space-sense of the blind is altogether different from that of a person who sees. Compare on this point the acute remarks of Dr. Th. Loewy (*Common Sensibles. Die Gemeinideen des Gesichts- und Tastsinnes nach Locke und Berkeley,* Leipzig, 1884), with whose results, however, I cannot agree. The circumstance that a man blind from birth does not, after being operated upon, in accordance with the experiment proposed by Molyneux, visually distinguish the cubes and spheres with which he is familiar from touch, proves to my mind nothing at all against Locke and nothing in favor of Berkeley and Diderot. Even persons who see recognize figures that are turned upside down only after much practice. The fact is that at the first moment of sight all the associations connected with the optical process, which may subserve its application intellectually, are wanting. A further point is that, when optical stimuli have been absent for a long period in early youth, the development of the central visual spheres may be arrested, or perhaps degeneration may even take place, as has been shown by Schnabel's beautiful observations ("Beiträge zur Lehre von der Schlechtsichtigkeit durch Nichtgebrauch der Augen," *Berichte des naturwissenschaftlich-medikalischen Vereins in Innsbruck,* Vol. XI., p. 32), and by Munk's experiments on new-born puppies (Berliner klinische Wochenschrift, 1877, No. 35). Even in the case of people who are not actually blind, the visual sphere may be so little developed that a special education is required to enable them to turn their sight-sensations to account. The case of the boy described by S. Heller, the Director of the Vienna Institute for the Blind (*Wiener klinische Wochenschrift,* 25th April 1901), is probably such a case of partial optical idiotism. It is only with great caution, therefore, that conclusions should be drawn from the behaviour, after operations, of those born blind. Chesselden, for instance, gave an account of an operation performed on a man blind from birth, who at first believed that everything he saw was in contact with his eyes; from this the false conclusion was drawn that the perception of the dimension of depth depends on extra-optical experiences. An accident put me in the way of understanding this phenomenon. Once when I was walking on a dark night in a district with which I was unfamiliar, I was all the time afraid of running up against a large black object. This turned out to be a hill several kilometres distant, which brought about this phenomenon through my being unable to fix and accommodate my sight, in much the same way as people who have been newly operated must be unable to do. If any one is not convinced by his own stereoscopy that the dimension of depth also is optically given, he is not likely to be convinced by the experiences of truncated people without arms and legs, such as Eva Lark and Kobelkoff (G. Hirth, *Energische Epigenesis*, 1898, p. 165).

All systems of space-sensation, however different they may be, are connected by a common associative link, the movements which they serve to guide. If Locke was wrong, how could the blind Saunderson have written a geometry intelligible to those who are not blind? Without doubt there are analogies between the sense of space given through sight and that given through touch. I have already mentioned something of this sort in discussing Soret's work, and many of these phenomena were known to the Aristotelian school. Thus in the Parva Naturalia we find mentioned the experiment by which a little ball is felt as double when touched by the index-finger and the middle finger placed across it. With me this experiment produces an even more striking effect when I cross my fingers and move them up and down a little stick; and when I take two parallel sticks, and arranging my fingers in this way. run them between them, I feel the two sticks as single. The analogy with seeing the single as double and seeing the double as single is here complete. But the differences also are so great that the man of normal sight finds it difficult to picture to himself a blind man's space-presentation, since he is always introducing his own visual presentations by way of interpretation. Even so acute a mind as Diderot's can fall on occasion into the strange error of denying that the blind can imagine space. Cf. Loeb's work; on tactual space, and Heller's *Studien zur Blindenpsychologie* (Leipzig, 1895). See Chapter IX.

[20] *Grundlinien der Lehre von den Bewegungsempfindungen*, Leipzig, Engelmann, 1875, p. 83.

[21] *Bewegungsempfindungen*, p. 124.

[22] Riehl, *Der philosophische Kritizismus*, Vol. II., p. 143.

[23] *Bewegungsempfindungen*, p. 26.

[24] Breuer, "Ueber die Funktion des Otolithen-Apparates," Pflüger's *Archiv*, Vol. XLVI., p. 195.

[25] *Bewegungsempfindungen*, p. 96, Experiment 2.

[26] *Bewegungsempfindungen*, pp. 116 sqq.

[27] *Bewegungsempfindungen*, p. 122.

[28] *Bewegungsempfindungen*, p. 85.

[29] As we all know, the most varied forms of the same impression are obtained in the midst of a number of railway trains some of which are in motion and some at rest. A short time ago, while making a steamboat excursion on the Elbe, I was astonished at getting the impression, just before landing, that the ship was stand-

ing still and that the whole landscape was moving towards it - an experience that will be readily understood from what follows.

[30] W. James, Principles of Psychology, II., pp. 512, sqq.

[31] Crum Brown, "On Normal Nystagmus".

[32] Such phenomena often make their appearance quite unsought. As my little daughter was once standing near a window, on a calm winter's day, during a heavy snowfall, she suddenly cried out that she and the whole house were rising upward.

[33] *Bewegungsempfindungen,* p. 85. For more recent experiments see A. von, Szily, "Bewegungsnachbild und Bewegungskontrast," *Zeitschrift für Psychologie und Physiologie der Sinnesorgane,* 1905. Vol. XXXVIII., p. 81.

[34] Karl Biehl, "Ueber intrakranielle Durchtrennung des Nervus vestibularis und deren Folgen," *Sitzungsberichte* of the Vienna Academy of Science, 1900.

[35] J. Breuer, "Studien über den Vestibularapparat," *Ibid.,* Vol. CXII., 1903.

[36] "The Otocyst of Decapod Crustacea, its Structure, Development and Functions," *Bulletin of the Museum of Comparative Zoology,* Harvard, 1900-1. (Quoted by Kreidl.)

[37] K. L. Schäfer, "Funktion und Funktionsentwicklung der Bogengänge, *Zeitschrift für Psychologie und Physiologie der Sinnesorgane,* 1894.

[38] Ach, "Ueber die Otolithenfunktion und Labyrinthtonus," Pflüger's *Archiv,* Vol. LXXXVI., 1900.

[39] William James, *American Journal of Otology,* 1887.

[40] A. Kreidl, "Beiträge zur Physiologie des Ohrlabyrinths auf Grund von Versuchen bei Taubstummen," Pflüger's *Archiv,* Vol. LI.

[41] J. Pollak, "Ueber den galvanischen Schwindelbei Taubstummen," etc., *Ibid.,* Vol. LIV.

[42] H. Mygind, "Ueber die pathologisch-anatomischen Veranderungen der Gehörorgane Taubstummer," *Ibid.,* Vol. XXV.

[43] Alexander and Kreidl, "Zur Physiologic des Labyrinthes der Tanzmaus," Pflüger's *Archiv,* I., II., III., Vols. LXXXII., LXXXVIII.

[44] Dreyfuss, "Experimentale Beiträge zu der Lehre von der nichtakustischen Funktion des Ohrlabyrinthes," Pflüger's *Archiv,* Vol. LXXXI.

[45] G. Alexander, "Zur Frage der phylogenetischen Ausbildung der Sinnesorgane," *Zeitschrift für Psychologie und Physiologie der Sinnesorgane,* Vol. XXXVIII.

[46] If I grasp a little bird in my hand, the bird will behave towards my hand exactly as a human being would towards a giant cuttlefish. - In watching a company of little children whose movements are largely unreflecting and unpractised, the hands and eyes remind one very strongly of polypoid creatures. Of course, such impressions do not afford solutions of scientific questions, but it is often very suggestive to abandon oneself to their influence.

[47] Compare Hering's opinion given in Hermann's *Handbuch der Physiologie,* Vol. III., Part I., p. 547. I do not wish to conceal the method by which I was led to my theory, although now the view represented by James, Münsterberg and Hering, as explained in Chapter VIII., seems to me preferable.

[48] This is the point to which reference was made earlier this chapter.

[49] This conception also removes a difficulty which I still felt in 1871, and to which I gave utterance in my lecture on "Symmetry" (Prague, Calve, 1872), - now

translated into English in my *Popular Scientific Lectures,* Chicago, 1894, - in the following words: "The possession of a sense for symmetry by persons who are one-eyed from birth is certainly an enigma. Yet the sense for symmetry, although originally acquired by the eyes, could not have been confined exclusively to the visual organs. By thousands of years of practice it must also have been implanted in other parts of the human organism, and cannot, therefore, be immediately eliminated on the loss of an eye." As a fact, *the symmetrical apparatus of innervation* remains, even when one eye is lost.

VIII. The Will

1.

IN what precedes, the phrase "the will" has often been used, and has always been intended merely to denote a generally recognized psychic phenomenon. I do not mean by the will any special psychical or metaphysical agent, nor do I assume a specific psychical causality. Rather, I am convinced, in company with the overwhelming majority of physiologists and modern psychologists, that the phenomena of volition must - to put it briefly, but in a way that everyone can understand - be explained by means of the physical forces of the organism alone. I would not lay any special emphasis on the fact that this is a matter of course, were it not that the remarks of many critics have shown the emphasis to be necessary.

The movements of lower animals, and, equally, the first movements of all new-born animals, are immediately set free by some stimulus; they follow the stimulus quite mechanically; they are reflex movements. Nor are such reflex movements absent in the later stages of the lives of higher animals, and when the occasion arises for us to observe them for the first time in ourselves, - for instance, the sinew-reflexes, - we are as much surprised by them as by any unexpected event in our environment. The behaviour of the young sparrow described above depends on reflex movements. The chick pecks quite mechanically at everything that it sees, just as the child grasps at everything that strikes its notice, and, on the other hand, withdraws its limbs from every unpleasant contact without any cooperation of the intellect. For there are fixed organic arrangements which determine the preservation of the organism. If we adopt Hering's views on living substance, according to which living substance strives towards the equilibrium of the antagonistic processes taking place in it, we shall be forced to ascribe to the elements of the organism themselves this tendency towards self-preservation or actual stability.

Sensational stimuli can be partly or wholly replaced by memory-images. All memory-traces that remain behind in the nervous system co-operate with the sensations to set free, to assist, to inhibit and to modify the reflexes. It is in this way that voluntary movement arises, since we can conceive voluntary movement, at any rate in principle, as reflex movement modified by memories, however far we may fall short of understanding it in detail. The child that has once burnt itself at a bright flame refrains in future from grasping at the flame, because the grasp-reflex is inhibited by the antagonistic avoidance-reflex which the memory of the pain sets free. The chick begins by pecking at everything; but soon, under the influence partly of inhibitory and partly of encouraging memories of taste, it exercises a choice. The gradual transition from reflex movement to voluntary action can be very prettily followed in the case of our sparrow. For the reflecting subject, the characteristic mark of voluntary action, as distinct from reflex movement, lies in the subject's recognition that the determining factor in voluntary action is his own presentations, which anticipate this action.

2.

The psychic processes that accompany voluntary action and voluntary movement have been admirably analysed by William James [1] and by Hugo Münsterberg. [2] It seems a simple and natural view to suppose that the actual movement is associated with the imagined movement in the same way as one presentation is associated with another. But as regards the sensations of the kind and amount of the movement, and of the amount of effort involved, which are connected with the execution of the movement, two opposite views are taken. According to one view, which is held by Bain, Wundt and Helmholtz, the innervation which flows to the muscles is itself felt. James and Münsterberg take a different view. They hold that all the kinaesthetic sensations that accompany a movement are peripherally excited by sensible elements in the skin, the muscles and the joints.

Against the hypothesis of the central origin of the kinaesthetic sensations we have to set, first and foremost, the observations on anaesthetic subjects, [3] who, when their sensations are cut off, are able to give no account of the passive movement of their limbs, although they are able to move their limbs under the guidance of the sense of sight. We feel the exertion of a faradised muscle just as much as in the case of a muscle that is voluntarily innervated. [4] The hypothesis of specific sensations of innervation is not required for the explanation of the phenomena, and, on the principle of economy, is consequently to be avoided. Finally, sensations of innervation are not directly observed. A special difficulty is constituted by certain optical phenomena, to which we shall return.

The law of association connects not merely processes that emerge into consciousness (presentations), but also the most diverse organic processes. The man who blushes readily when he is embarrassed, whose hands sweat readily, etc., generally observes these processes taking place in himself the moment he is reminded of them. For purposes of study Newton [5] acquired a dazzling after-image by gazing at the sun; this image disappeared, but during a period of several months, although he remained for several days in the dark, it always returned with full sensational intensity the moment he re-minded himself of it. It was only by diverting his attention by a long-continued and violent psychic effort that he was able to get rid of this trou-blesome phenomenon. Boyle narrates a similar observation in his book on colors. When brought into connexion with these facts, the association of mo-tor processes with presentations ceases to appear strange.

3.

An apoplectic stroke which I experienced in 1898, without its in the least affecting my consciousness, has made me personally familiar with part of the facts now under consideration. I was in a railway train, when I suddenly ob-served, with no consciousness of anything else being wrong, that my right arm and leg were completely paralysed; the paralysis was intermittent, so that from time to time I was able to move again in an apparently quite nor-mal way. After some hours it became continuous and permanent, and there also set in an affection of the right facial muscle, which prevented me from speaking except in a low tone and with some difficulty, I can only describe my condition during the period of complete paralysis by saying that when I formed the intention of moving my limbs I felt no effort, but that it was abso-lutely impossible for me to bring my will to the point of executing the move-ment. On the other hand, during the phases of imperfect paralysis, and dur-ing the period of convalescence, my arm and leg seemed to me enormous burdens which I could only lift with the greatest effort. It seems to me plau-sible to suppose that this was caused by the energetic innervation of other muscle-groups in addition to the muscles of the paralysed extremities. [6] The paralysed limbs retained their sensibility completely, except for one place on the thigh, and thus I was enabled to be aware of their position and of their passive movements. I found that the reflex excitability of the para-lysed limbs was enormously heightened; this expressed itself particularly in violent jerks on the slightest alarm. Optical and tactual motor-images per-sisted in my memory. Very often during the day I formed the intention to do something with my right hand, and had to think of the impossibility of doing it. To the same source are to be referred the vivid dreams which I had of playing the piano and writing, accompanied by astonishment at the ease with which I wrote and played, and followed by bitter disappointment on awak-ing. Motor hallucinations also occurred. I often thought that I felt my para-lysed hand opening and shutting, and at the same time the total movement seemed to be hampered as if by a loose, but stiff glove. But I had only to look to convince myself that there was not the slightest movement. Over the flex-ors of this hand I have acquired a very slight control, but over the extensors none at all.

Since the sensibility of the hand is preserved, while the power of voluntary movement is lacking, I do not know how to explain the illusion of movement properly, even on the new theory. The muscles that are withdrawn from the influence of the will react now to the most diverse stimuli, so that the hand is sometimes extended and sometimes clenched. Strong tastes of different qualities seem to act as stimuli to different extents on different muscles of my paralysed hand. Water impregnated with sulphate of magnesia, for example, excites involuntary movements of tension in the thumb and the two fingers next to it.

4.

The theory of James and Münsterberg fits these facts, as I think, without any straining, and we ought therefore to consider it as correct in essentials. The innervation is not felt, but the consequences of the innervation set up new peripheral sensible stimuli, which are connected with the execution of the movement. There are, however, some difficulties which prevent me from believing that this view which was originally my own, [7] completely explains the facts.

One would think that the central process which conditions the mere presentation of a movement must differ in some respect from the process which also releases an actual movement. To be sure, the strength of the process, the absence of antagonistic processes, and the extent to which the innervation-centres are charged, may be partial determinants; but still it is scarcely possible to deny that further explanation is required. In particular, the difference in behaviour of the muscles of the eye and the other muscles that can be excited at will needs closer investigation. Most muscles have variable amounts of work to perform, and it is of practical importance for us to know these amounts approximately. The work done by the eye-muscles, on the other hand, is small, and is always exactly connected with the alteration in position of the eyes; this latter alone is of optical importance, while the work" as such is matter of indifference. This may be the reason why the kinaesthetic sensations play such a much greater part in the case of the muscles of the extremities.

5.

Hering [8] has shown how small is the importance of the sensations proceeding from the muscles of the eye. Usually we scarcely attend at all to the movements of our eyes, and the position of objects in space remains uninfluenced by these movements. If one imagines two spherical surfaces, covered with movable retinae, and remaining fixed in space while the retinae revolve, a superficial consideration might even induce us to believe that the space-values of the objects seen would only be determined by the two positions of reflexion on the fixed spheres. But the facts mentioned above (Ch. VII.), compel us to separate these space-values into two components, one of which depends on the co-ordinates of the point of reflexion on the retina, and the other on the co-ordinates of the point of vision, which components undergo mutually compensating alterations corresponding to voluntary alterations of the point of vision. [9] Now, if we refuse to admit a sensation of innervation, and deny the importance of the peripherally excited kinaesthetic sensations of the eye-muscles, the only remaining alternative is that adopted by Hering, - namely to regard the position of attention as determined by a definite psycho-physical process, which at the same time is the physical factor that releases the corresponding innervation of the eye-muscles. [10] But this process is still a central process, and "attention" remains scarcely different from "the will to see." In this way I might still hold in essentials to my expression on p. 130 above; for which of the series of processes excited from and proceeding from the centre is the one that enters into sensation, is a many-sided problem which can for the present remain unsolved.

6.

In conformity with what precedes, we might replace, in the explanation attempted at the end of Ch. VII., the two antagonistic innervations by two antagonistic processes of attention, one excited by the sensible stimulus, and the other a central process. I cannot give my assent to the explanation proposed by James 1 of the phenomena connected with paralysis of the muscles of the eye, since this explanation seems, formally at any rate, to drift towards the doubtful waters of "unconscious inferences." In the case under discussion we are concerned with sensations and not with the results of reflexions.

The function of the muscles of the eyes is merely to ensure our orientation in space; that of the muscles of the limbs is principally the performance of mechanical work. We thus have here two extreme cases, between which there will lie many middle terms. When we see the newly-hatched chick pecking and hitting its mark accurately, it is easy to believe that the muscles of its neck and head to some extent perform a similar office to that of its eye-muscles, and act as an apparatus of spatial orientation. Probably the jerking movements of the head which take place in birds when they walk forward, are executed, like nystagmic movements of the head under rotation, in the interests of orientation. The muscles of the extremities also cannot be entirely without analogy to the eye-muscles. Otherwise how could we understand the blind man's tactual presentation of space? For it is not easy to combine a nativistic theory of visual space with an empirical theory of tactual space. [12]

[1] W. James, *Principles of Psychology*, Vol. II., pp. 486 sqq.
[2] H. Münsterberg, *Die Willenshandlung*, Freiburg im Breisgau, 1888.
[3] W. James, *op. cit.*, Vol. II., p. 489.
[4] W. James, *op. cit.*, Vol. II., p. 502.
[5] King's *Life of Locke*, 1830, Vol. I., p. 404; Brewster, *Memoirs of Newton*, 1855, Vol. I., p. 236.
[6] W. James, *op. cit.*, Vol. II., p. 503.
[7] Before the phenomena connected with paralysis of the eye-muscles were known to me, *i.e.*, before 1863.

[8] Hering in Hermann's *Handbuch der Physiologie,* Vol. III., I, 547. Cf. also Hillebrand, "Verhältnis der Akkommodation und Konvergenz zur Tiefenlokalisation," *Zeitschrift für Psychologie und Physiologie der Sinnesorgane,* Vol. VII., pp.97, sqq.

[9] Cf. Ch. VI. above; Hering, *op. cit.* pp. 533, 534. I am now unable to decide whether the view that the alteration of the space-values is completed immediately with the change in attention can be brought into harmony with the fact mentioned in Ch. VII. above.

[10] Hering, *op. cit.,* pp. 547, 548.

[11] W. James, *op. cit.,* Vol. II., p. 506.

[12] Cf. Ch. VII.

IX. Biologico-Teleological Considerations as to Space

[1]

1.

WE have already repeatedly had occasion to notice how very different the system of our space-sensations - our physiological space, if we may use the expression - is from geometrical (by which is here meant Euclidean) space. This is true, not only as regards visual space, but also as regards the blind man's tactual space in comparison with geometrical space. Geometrical space is of the same nature everywhere and in all directions, it is unlimited and (in Riemann's sense) infinite. Visual space is bounded and finite, and, what is more, its extension is different in different directions, as a glance at the flattened "vault of heaven" teaches us. Bodies shrink when they are removed to a distance; when they are brought near they are enlarged: in these features visual space resembles many constructions of the metageometricians rather than Euclidean space. The difference between "above" and "below," between "before" and "behind," and also, strictly speaking, between "right and left," is common to tactual space and visual space. In geometrical space there are no such differences. For man, and for animals of similar structure to man, physiological space is related to geometrical very much as a "triclinal" is to a "tesseral" medium. This is true for men and animals, so long as they are not endowed with freedom of movement and of orientation. When mobility is added, physiological space approximates to Euclidean, though without completely attaining to the simplicity of the properties of Euclidean space. Three-dimensional multiplicity and continuity are common to geometrical and to physiological space. To the continuous movement of a point A in geometrical space, corresponds a continuous movement of a point A' in physiological space. We have only to remember the difficulties which the doctrine of the antipodes had to overcome, to see how geometrical space-presentations can be disturbed by physiological. Even our most abstract geometry does not employ purely metrical notions, but uses also such physiological conceptions as direction, sense, right, left, etc.

In order to keep the physiological and geometrical factors completely apart, we have to reflect that our space-sensations are determined by the dependence of the elements which we have called *A B C*...upon the elements of our body *K L M*...but that geometrical conceptions are formed by means of the spatial comparison of bodies by the relations of the *A B C's*... to one another.

2.

If we consider the spatial sensations not as isolated phenomena, but in their biological connexion, in their biological function, they become easier to understand, at any rate teleologically. As soon as an organ or a system of organs is stimulated, movements take place as reflex reactions; these movements are generally purposive, and may be defensive or offensive, according to the nature of the stimulus. For example, different places on the skin of a frog may be successively stimulated by dropping acid on them. To each excitation the frog will reply by a specific defensive movement according to the spot stimulated. The stimulation of places on the retinae sets free the equally specific reflex of snapping. That is to say, alterations that make their entrance into the organism by different paths are reproduced externally also by different paths on the animal's environment. Now, let us suppose that, in complicated conditions of life, such reactions can also arise spontaneously by memory, that is to say, on a slight impulse, and that they can be modified by memories; then traces, corresponding to the nature of the stimulus and to the stimulated organs, must remain behind in the memory. As our observation of ourselves teaches us, we not only recognize identity of quality in the stimulus of burning, whatever be the spot burnt, but at the same time we distinguish between the spots stimulated. We must therefore suppose that there is attached to the qualitatively identical sensation an element of difference, which depends on the specific nature of the elementary organ stimulated, on the spot stimulated, or, to use Hering's expression, on the "place of attention." It is precisely in the perception of space that the intimate mutual biological adaptation of a multiplicity of connected elementary organs is displayed with peculiar clearness.

3.

Let us assume only one kind of element of consciousness, namely sensations. In so far as we have spatial perceptions, these depend, according to our theory, on sensations. What is the nature of these sensations, and what organs are active in connexion with them, we must leave an open question. We have to imagine a system of elementary organs of common embryological origin as being naturally so arranged that neighbouring elements display the greatest ontological affinity, and that this affinity decreases as the segregation of the elements from one another increases. The organic sensation, which alone depends upon the individuality of the organ, and which varies with the variation in the degree of affinity, will correspond to the sensation of space, and from this we distinguish the sensations which depend on the quality of the stimulus as *specific sensations*. Organic sensations and specific sensations can only appear concomitantly. [2] But, over against the varying specific sensations, the unchanging organic sensations presently constitute a fixed register, in which the former are arranged. We are here only making as to the elementary organs similar presuppositions such as we should find natural in the case of separated individuals of the same descent but of different degrees of affinity.

4.

The perception of space has arisen from biological necessities, and can best be understood by reference to these necessities. An endless system of space-sensations would not only be purposeless for the organism, but would also be physically and physiologically impossible. Space-sensations which should be orientated with reference to the body would also be valueless. It is also advantageous that, in visual space, the sensation-indices for nearer objects, which are biologically the most important, should be more sharply graded, while as regards more remote, and consequently less important, objects, the limited supply of indices is used economically. This relation, again, is the only one physically possible.

The following considerations enable us to understand the motor organization of the visual apparatus. The greater clearness and the finer distinctions that exist at a given spot on the retina of the eye of a vertebrate are an economic arrangement. By this means a movement of the eyes that follows a change of attention is recognized as advantageous, just as the influence of a voluntary movement of the eyes on the space-sensation caused by objects at rest is recognized as disadvantageous, if it is misleading. Nevertheless the displacement of images on the retina, which itself remains at rest, is biologically necessary, in order to enable us to perceive moving objects with our eyes at rest: the only case in which it is unnecessary for the organism, is the very rare one when it is required to perceive an object at rest when the eye is moved by some cause that does not emerge into consciousness, such as an external mechanical force or a twitching of the muscles. The only solution that provides for all the foregoing requirements is that, when the eye is moved voluntarily, the displacement of images on the retina corresponding to this movement should be compensated by the voluntary movement in respect of space-value. But from this it follows, that when the eye is kept still, the mere intention to move the eye must cause objects at rest to undergo some displacement in visual space. By means of a suitable experiment (p. 128 above) the existence of the second also of the two mutually compensatory components is directly proved. These organic arrangements are ultimately the reason why under peculiar circumstance, when our eyes are at rest, objects at rest appear to move and the space-values to fluctuate, why we see bodies in motion, which nevertheless do not change their position relative to our body, and neither move farther away nor come nearer. What seems paradoxical under these peculiar circumstances, is, under ordinary circumstances (those of spontaneous locomotion) of the highest biological importance.

The relations of tactual space are, apart from certain peculiarities, similar to those of visual space. The sense of touch is not a long-range sense; there is no perspectival shrinkage and enlargement of tactual objects. But otherwise the phenomena which we find here are akin to those of vision. The finger-tips correspond to the macula lutea. We can tell the difference perfectly well between passing our finger-tips over a motionless object and the movement of an object over our motionless fingertips. Analogous paradoxical phenomena connected with rotatory vertigo appear here also. They were known to Purkinje.

5.

Biological considerations of a general nature force us to conceive optical and tactual space as homogeneous. A newly-hatched chick notices a small object and at once both looks and pecks at it. The stimulus excites a certain tract of the sense-organ and of the central organ, by means of which both the looking movement of the eye-muscles and the pecking movement of the muscles of the head and neck are released perfectly automatically. The excitement of one and the same nerve-tract, which, on the one hand, is determined by the geometrical position of the physical stimulus, must, on the other hand, be regarded as the foundation of the sensation of space. A child that notices a glittering object, then looks at it and grasps at it, behaves in the same way as the chick. In addition to optical stimuli there are others - acoustic stimuli, stimuli of heat and smell - which can also release movements of grasping or of defence; and these of course are operative with blind people also. Again, the same places of stimulus and the same sensations of space will correspond to the same movements. In general, the stimuli that excite a blind person are only confined to a narrower sphere and are less clearly and definitely localized. The system of his space-sensations will consequently be rather poorer and more blurred than that of a normal person, and, in the absence of a special education, will even remain so. Consider, for example, a blind man who tries to keep off a wasp that is buzzing round him.

Accordingly as an object stimulates me to turn my gaze upon it, or stimulates me to seize it, the tracts of the central organ that come into play must be partially different, although adjacent. If the two stimuli take place at once, the tract involved is of course larger. On biological grounds we should expect that the space-sensations connected with different senses, which, though not identical, are closely akin to one another, would be linked together by means of the movements which they induce, - movements which are directed to the preservation of the organism, - would merge into one another by way of association, and would mutually support one another; and this is in fact the case.

But this conclusion does not exhaust the phenomena with which we are concerned. A chick may look at an object or may peck at it; it may also be determined by the stimulus to turn towards it and to run up to it. Exactly similar is the behaviour of a child that crawls towards some goal, and then one day stands up and runs a few steps towards the goal. All these cases pass gradually over into one another, and we have to think of them all as homogeneous. Probably there are always certain parts of the brain which! when stimulated in a relatively simple manner, determine space-sensations on the one hand, and on the other hand release automatic movements which are sometimes extremely complicated. Optical, thermal, acoustic, chemical and galvanic stimuli can excite a great deal of locomotion and change of orientation, and these effects can be produced even in animals that are blind, either originally or by degeneration.

6.

When we observe a millipede (*Julus*) crawling regularly on its way, we cannot resist the thought that a uniform current of stimulus proceeds from some one of the insect's organs, and that the motor organs of the successive segments of the body reply to this current with rhythmic automatic movements. The longitudinal wave, which seems to pass along the insect's rows of feet with machinelike regularity, arises from the difference in phase of the segments behind as compared with those in front. We should expect to find analogous processes in the more highly organized animals, and in fact such processes occur. I need only refer to the phenomena connected with stimulations of the labyrinth, for instance to the well-known nystagmic movements of the eyes, which are released under active and passive rotation. Now, if there are organs, as in the case of the millipede, by the simple stimulation of which, the complicated movements of a definite kind of locomotion are induced, then we may regard this simple stimulation, in the case where it is present to consciousness, as the *will* to the locomotion in question, or as the *attention* to the locomotion, which automatically draws the locomotion after it. At the same time, we recognize that it is a necessity for the organism to feel the effect of the locomotion in a correspondingly simple manner. And, in fact, objects of sight and touch do appear with varying and fluctuating space-values, instead of their values being stable. Even if we exclude all sensations of sight and touch as completely as possible, there still remain sensations of acceleration, which evoke, by way of association, the images of the various space-values with which they have often been connected. Between the first and last members of the process lie the sensations of movement in the extremities, which, however, are usually only fully present to consciousness when some obstacle arises and makes a modification of the movement necessary.

The man who is motionless as a whole is only aware of space-sensations which are limited, locally individualized, and orientated with reference to his own body; but the sensations which arise on occasion of locomotion and change of orientation possess the character of regularity and inexhaustibility. All these experiences are required as a basis for the construction of a conception of space approximately similar to Euclidean space. Apart from the fact that the first set of experiences only gives us agreements and differences, but no magnitudes and no metrical determinations, the latter set does not attain absolute uniformity on account of the obstacles in the way of permanent and free disorientation in respect of the vertical.

7.

For the animal organism, the relations between the parts of its own body are, first of all, of the highest importance. An alien object only acquires value by standing in relation to parts of the animal's own body. In the lowest organisms the sensations, including the space-sensations, are sufficient to secure adaptation to primitive conditions of life. But as these conditions become more complicated they force on the development of the intellect. Then the mutual relations of those functional complexes of elements (sensations) which we call bodies, acquire an indirect interest. Geometry arises from the spatial comparison of bodies with one another.

Our understanding of the development of geometry may be assisted by the observation that our immediate interest is connected, not with spatial properties alone, but with the whole permanent complex of material properties which is important for the satisfaction of our needs. But the forms, positions, distances, and extensions of bodies are decisive for the mode and the quantity of the satisfaction of needs. Mere perception by itself (estimation, ocular measurement, and memory) proves to be too much under the influence of physiological circumstances that are not easily controlled, for us to build securely upon it, when the question is one of judging accurately the spatial relations of bodies to one another. We are therefore compelled to look to the bodies themselves for trustworthy indications.

Everyday experience brings home to us the permanence of bodies. Under ordinary circumstances this permanence extends also to particular qualities, such as color, shape, and size. We become acquainted with rigid bodies, which, in spite of their mobility in space, always produce the same space-sensations as soon as they are brought into a definite relation with our body and are seen and handled. Such bodies display spatial substantiality; l they remain spatially constant and identical. One rigid body A may be immediately or mediately superimposed spatially on another rigid body B, or on parts of it, and this relation remains always and everywhere the same. We then say B is measured by A. When bodies are compared with one another in this way, the question is no longer one as to the kind of space-sensation involved; rather, we have a judgment as to their identity under similar circumstances, and this judgment is formed with great accuracy and certainty. Variations in the results of measurement are, in fact, negligible in comparison with the element of error involved in immediate spatial judgments as to juxtaposed or successive bodies, and it is in this fact that the advantage and the rational justification of the process of measurement consist. Instead of the individual hands and feet, which everyone carries about with him without noticing any appreciable spatial change in them, a universally accessible standard of measure is soon chosen, which, by fulfilling in a high degree the condition of immutability, ushers in an era of greater precision.=

8.

The object of all geometrical problems is to establish a numerical correspondence between spaces that it is required to ascertain and known homogeneous bodies. Empty vessels for the measurement of fluids, or of compact aggregates of almost exactly similar bodies, are probably the oldest measures. The volume of a body, - the aggregate of the positions occupied by its matter, - which we instinctively represent to ourselves when we look at or grasp somebody with which we are familiar, comes to be considered as a quantum of material properties that satisfy our needs, and constitutes, as such, an object of dispute. Indeed, originally the measurement of a surface is undertaken solely with the object of ascertaining the amount of the homogeneous closely juxtaposed bodies covering the surface. Measurement of length, - that is to say, enumeration by means of similar pieces of string or chain, - determines the minimum volume that can be interpolated in one and only way between two points, or two very small bodies. If in this process we abstract from one or two dimensions of the bodies used as measures, or, again, if we suppose these bodies to be everywhere constant, but as small as we like to choose them, we arrive at the idealized representations of geometry.

9.

Our intuition of space is enriched by experimenting with material objects, owing to the fact that metrical experiences, which spatial intuition would not be able to acquire by itself, are connected with these objects. Thus we become acquainted with the metrical properties of forms with which we have long been familiar, such as the straight line, the plane, and the circle. Again it is experience, as history testifies, which has first led to the knowledge of certain geometrical propositions, by showing that, if an object has certain dimensions, certain other dimensions of it were thereby determined. Scientific geometry set itself the economical task of ascertaining the dependence of dimensions on one another, of avoiding superfluous measurements, and of discovering the simplest geometrical facts from which the remaining facts would follow as logical consequences. For this purpose, since we have no mental control over nature, but only over our own simple logical constructions, our ultimate geometrical experiences had to be conceptually idealized. Henceforth there was no obstacle in the way of the discovery of geometrical propositions by a kind of "thought-experiment," - by advancing along the road of mental visualization, and thinking of these representations as connected with the idealized geometrical experiences. The procedure throughout is analogous to that of all the natural sciences. But the ultimate experiences of geometry are reduced to so small a minimum that is only too easy to overlook them altogether. We imagine bodies as moving over the shadows or ghosts of bodies, and we cling mentally to the notion that their measurements, if they were taken, would not be altered in the process. Physical bodies harmonize with the results in so far as they are sufficient for the presuppositions.

Intuition, physical experiences, and conceptual idealization, are, therefore, the three co-operating factors in scientific geometry. The wide divergence in the views of different investigators as to the nature of geometry is due to over - or under - estimation of one or the other factor. The only possible foundation for a correct view is the precise separation of the part played by each of these factors in the building up of geometry. For instance, our anatomically symmetrical motor organization, which has been acquired for purposes of rapid locomotion, causes our intuition to make the two halves of a spatially symmetrical construction appear to us as equivalent; but this is by no means true from a physico-geometrical point of view, since they cannot be brought into congruence. Physically they are no more equivalent than a movement is to an opposite movement, or a rotation to a rotation in the opposite direction. Kant's paradoxes on this subject depend on an inadequate separation of the various factors involved.

[1] This subject cannot be treated in detail here. I may refer to my articles in *The Monist,* of which the first appeared in April 1901, the second in July 1902, and the third in October 1903. The physiological considerations outlined here are partly related to Wlassak's views as stated in his paper, "Ueber die statischen Funktionen des Ohrlabyrinthes" *Vierteljahrsschrift für wissenschaftliche Philosophie,* Vol. XVII., I, p. 28), except that I assume, not one, but two, reactions to the stimuli in question. Cf. also the passages cited above from Hering, and W. James, *Psychology,* Vol. II., pp. 134, sqq. Cf. also my *Erkenntnis und Irrtum,* 1905, pp. 331-414, 426-440.

[2] Similarly the internal organs are only sensed and localized when some disturbance of their equilibrium takes place.

[3] It is certain that this view has been privately held by innumerable geometricians. It comes out clearly in the whole arrangement of Euclid's geometry, and is still clearer in Leibniz, particularly in his "geometrical characteristic." But Helmholtz was the first to discuss it openly.

X. The Relations of the Sight-Sensations to One Another and to Other Psychical Elements

1.

IN normal psychical life, sight-sensations do not make their appearance alone, but are accompanied by other sensations. We do not see optical images in an optical space, but we perceive the bodies round about us with their many and varied sensible qualities. Deliberate analysis is needed to single out the sight-sensations from these complexes. But even the total perceptions themselves are almost invariably accompanied by thoughts, wishes, and impulses. By sensations are excited, in animals, the movements of adaptation demanded by their conditions of life. If these conditions are simple, altering but little and slowly, immediate sensory excitation is sufficient. Higher intellectual development is unnecessary. But the case is different where the conditions of life are intricate and variable. Here so simple a mechanism of adaptation cannot develop, still less would it lead to the accomplishment of the required ends.

Lower species devour everything that comes in their way and that excites the proper stimulus. A more highly developed animal must seek its food at risks to itself; when found, must seize it at the right spot, or capture it by cunning, and cautiously test its character. Long trains of varied memories must pass before its mind before one is sufficiently strong to outweigh the antagonistic considerations and to excite the appropriate movement. Here, therefore, a sum of associated remembrances (or experiences) coincidently determining the adaptive movements, must accompany and confront the sensations. In this consists the intellect.

In the young of higher animals under complex conditions of life, the complexes of sensations necessary to excite adaptive movements are frequently of a very complicated nature. The sucking of young mammals, and the behaviour of the young sparrow described in Ch. IV. are good examples of this. With the development of intelligence, the parts of these complexes necessary to produce the excitation constantly diminish, and the sensations are more and more supplemented and replaced by the intellect, as may be daily observed in children and adolescent animals.

In a note to the edition of 1886 I uttered a warning against the tendency, which was still widespread at that time, to over-estimate the intelligence of the lower animals. My view was based solely on occasional observations on the machine-like movement of beetles, the flight of moths towards the light, etc. Subsequently the important works of J. Loeb appeared, and provided a solid experimental basis for this view.

At the present moment (1906) the psychology of the lower animals has again become the field of much controversy. While A. Bethe [1] advocates an extreme reflex-theory, based on ingenious and interesting experiments on ants and bees, according to which these insects are to be regarded as Cartesian machines, other careful critical observers, such as E. Wasmann, [2] H. von Buttel-Reepen, [3] and A. Forel [4] ascribe to the same insects a high degree of psychic development. The psychology of the higher animals also has lately become the object of general interest. The writings of Theodor Zell, which are intended principally for the general public, are full of excellent observation and felicitous insight, and seem to hit with great caution the proper mean between over-estimating and under estimating the animals of which they treat.

Anyone who has studied physiology, or even anyone who can appreciate the work of F. Goltz, knows the very important part played by reflexes in preserving the life of all animal organisms, even of the human organism, which is the most highly developed of all. To anyone, again, who has observed the striking way in which the influence exercised upon the biological reactions by a memory that registers the experiences of the individual, decreases with the simplification of the organism, it will naturally occur to try whether and to what extent the behaviour of simpler organisms can be explained solely by reference to reflexes. [5] It is not, indeed, probable that there exist any animal organisms entirely devoid of memory and endowed with reflexes absolutely incapable of modification, since it is scarcely possible to draw a sharp line between the acquisitions of the species and the acquisitions of the individual. [6] Still, I should consider such an attempt as well worth making, although a critical analysis of the result would be still more valuable.

I hope that we shall still learn a great deal for our own psychology, not only from our children, but also from "our younger brothers" the animals. But in order to understand why man is psychically so much more than the cleverest animal, it will be sufficient to reflect on the acquisitions which the individual and the species have made in the atmosphere of a social culture extending over many thousand years.

2.

Representation by images and ideas, therefore, has to supply the place of sensations, where the latter are imperfect, and to carry to their issue the processes initially determined by sensations alone. But in normal life, representation cannot permanently supplant sensation, where this is at all present, except with the greatest danger to the organism. As a fact, there is, in normal psychical life, a marked difference between the two species of psychical elements. I see a blackboard before me. I can, with the greatest vividness, represent to myself on this blackboard, either a hexagon drawn in clear, white lines, or a colored figure. But, pathological cases apart, I always distinguish what I see and what I represent to myself. In the transition to mental imagery, I am aware that my attention is turned from my eyes, and directed elsewhere. In consequence of this attention, the spot seen upon the blackboard and the one represented to myself as situated in the same place differ as by a fourth co-ordinate. It would not be a complete description of the facts to say that the image is superimposed on the object as the images reflected in a transparent plate of glass are superimposed on the bodies seen through it. On the contrary, what is represented seems to me to be supplemented by a qualitatively different and opposite sensational stimulus, which stimulus it in its turn sometimes supplants. We are confronted here, for the time being, with a psychological fact, the physiological explanation of which will sometime undoubtedly be discovered.

It is natural to suppose that, when mental images occur, the interaction of the organs of the nervous system causes the repetition of organic processes partially identical with those which were determined by the physical stimulus on occasion of the corresponding sensations. Images are normally distinguished from sensations by being less intense, and above all by their instability. When I draw a geometrical figure in imagination, it is as if the lines faded immediately after they are drawn, as soon as my attention is directed to ether lines: when one comes back to them they have vanished, and must be reproduced over again. This is the principal reason of the advantage in point of convenience which an actual material geometrical drawing possesses over a merely imagined one. It is easy enough to hold firmly before the mind a small number of lines, for instance an arc of a circle with the angles at the centre and circumference and a pair of coincident or intersecting sides; but if in this case we proceed to add the diameter drawn through the apex of the angle at the circumference, it at once becomes more difficult to deduce in imagination the relative sizes of the angles, without continually renewing and completing the figure. The power of replacing the figure with ease and rapidity, is, however, enormously increased by practice. When I was studying the geometry of Steiner and Von Staudt, I was able to do this to a much greater extent than I can now.

Where the development of intelligence has reached a high point, such as is presented now in the complex conditions of human life, mental images may frequently absorb the whole of attention, so that events in the neighborhood of the reflecting person are not noticed, and questions addressed to him are not heard; a state which persons unused to it are wont to call absentmindedness, although it might with more appropriateness be called present-mindedness. If the person in question is disturbed in such a state, he has a very distinct sensation of the labor involved in the transference of his attention.

3.

It is well to note this sharp division between images and sensations, as it is an excellent safeguard against carelessness in psychological explanations of sense-phenomena. The well-known theory of "unconscious inferences" would never have reached its present extended development if more heed had been paid to this circumstance.

The organ, of which the states determine images, can provisionally be conceived as one which, in a diminished degree, is capable of all the specific energies of the sense and motor organs, so that the specific energy now of one, now of another, sense-organ can play upon it, according to the nature and direction of its attention for the time being. Such an organ is eminently qualified to effect physiological relations between the different energies. As is shewn by experiments with animals whose cerebrum has been removed, there are probably, in addition to the organ of representation, a number of other analogous organs of mediation, which are less intimately connected with the cerebrum, and whose processes consequently do not appear in consciousness.

That wealth of representative life with which we are personally acquainted from self-observation, doubtless made its first appearance with man. But the beginnings of this expression of life, in which nothing but the relations of the various parts of the organism to one another is manifested, go back with no less certainty to quite primitive stages in the animal scale. But the parts of single organs must also, in virtue of their reciprocal tension, stand to one another in a relation analogous to that in which the parts of the organism as a whole stand to one another. The two retinae, with their motor mechanism of accommodation and of luminous adjustment, dependent on light-sensations, afford a very clear and familiar example of such a relation. Physiological experiment and simple self-observation teach us that such an organ has its own purposive habits, its own peculiar memory, one might almost say its own intelligence.

4.

The most instructive observations in this connexion are probably those of Johannes Müller, collected in his admirable work on "The Phantasms of Sight" (*Ueber die phantastischen Gesichtserscheinungen,* Coblenz, 1826). The sightphantasms observed by Müller and others in the waking state are entirely withdrawn from the influence of either the will or the reason. They are independent phenomena, essentially connected with the sense-organs, and characterized by complete visual objectivity. They are veritable imaginationand memory-phenomena of sense. Müller considers that the free individual existence of hallucinations is a part of the life of the organism, and that it cannot be brought under the so-called laws of association, in which he indicates that he does not believe. It seems to me that the continuous alteration of the phantasms, as described by him, is no evidence against the laws of association. These processes can, on the contrary, be conceived as recollections of slow perspectival changes in visual images. The element of desultoriness in the common connexion of a train of representations by way of association, only comes in when sometimes one, and sometimes another, department of sensation begins to be involved (see Chapter Eleven.)

Those processes which (according to Müller) are normally induced in the "visual substance" by excitations of the retina, and which condition the act of seeing, may also, under exceptional conditions, be spontaneously produced in the visual substance without excitation of the retina, and thus become the source of phantasms or hallucinations. We speak of *sense-memory* when the phantasms are closely allied in character to objects seen before, of *hallucinations* when the phantasms arise more freely and independently. But no sharp distinction between the two cases can be maintained.

I am acquainted with all manner of sight-phantasms from my own experience. The mingling of phantasms with objects indistinctly seen, the latter being partly supplanted, is probably the most common case. In my own case, these phenomena are particularly vivid after a tiring night's journey in the train. Rocks and trees then assume the most fantastic shapes. Years ago, while engrossed with the study of pulse-tracings and sphygmography, the fine white curves on the dark background often came up before my eyes, in the evening or in the dim light of day, with the full semblance of reality and objectivity. Later also, during miscellaneous work in physics, I witnessed analogous phenomena of "sense-memory." More rarely, images of things which I have never seen before, have appeared before my eyes in the daytime. Thus, years ago, on a number of successive days, a bright red capillary net (similar to a so-called enchanted net) shone out upon the book in which I was reading, or on my writing paper, although I had never been occupied with forms of this sort. The sight of bright-colored changing carpet patterns before falling asleep was very familiar to me in my youth; the phenomenon will still make its appearance if I fix my attention on it. One of my children, likewise, often used to tell me that he "saw flowers" before falling asleep. Less often, I see in the evening, before falling asleep, various human figures, which alter without the action of my will. On a single occasion I attempted successfully to change a human face into a fleshless skull; this solitary instance may, however, be an accident. It has often happened to me that, on awaking in a dark room, the images of my latest dreams remained present in vivid colors and in abundant light. A peculiar phenomenon, which has for some years frequently occurred with me, is the following. I awake and lie motionless with my eyes closed. Before me I see the bedspread with all its little folds, and upon it, motionless and unchanging, my hands in all their details. If I open my eyes, either it is quite dark, or it is light, but the bed-spread and my hands lie quite differently from the manner in which they appeared to me. This is a remarkably fixed and persistent phantasm with me, such as I have not observed under other conditions. As regards this image, I think that I notice that all its parts, even those that are widely separated, appear with equal distinctness in a way which for obvious reasons is impossible in the case of anything objectively seen.

When I was young I used frequently to have very vivid acoustic, and particularly musical, hallucinations on waking up; they have, however, become extremely rare and faint since my interest in music has decreased. But perhaps the interest in music is itself a secondary effect, rather than a cause.

When we withdraw the retina from the influence of outward excitations, and turn the attention to the field of vision alone, traces of phantasms are almost always present. Indeed, they make their appearance when the outer excitations are merely weak and indistinct, in a half-light, or when we look at a surface covered with dim, blurred spots, such as a cloud, or a grey wall. The figures which we then seem to see, provided they are not produced by a direct act of attention in selecting and combining distinctly seen spots, are certainly not products of representation, but constitute, at least in part, spontaneous phantasms, which, for the time being and at some points, take forcible precedence over the retinal excitation. In these cases expectation seems to be favorable to the occurrence of the phantasms. When I have been looking for interference-bands I have very often thought that I could clearly detect the first dull traces of them in the field of vision, when the progress of the experiment has convinced me that I was certainly deluded. Over and over again, in a half-light, I have thought that I could distinctly see a jet of water that I was expecting to come out of an india-rubber tube, and have had to touch with my finger to convince myself of my mistake. Such weak phantasms seem as a rule to yield readily to the influence of the intellect, whereas the intellect is unable to produce any effect on strong phantasms with vivid colors. The former are more akin to representations, the latter to sensations.

These weak phantasms, which are sometimes overpowered by sensations, are sometimes in equilibrium with them, and sometimes replace them, suggest the possibility of comparing the strength of phantasms with the strength of sensations. Scripture has carried out this idea. He takes an observer who thinks that he sees a colored cross, which is really non-existent, and then causes to appear in his field of vision a real line of intensity increasing from zero upwards and drawn in a direction which is not known beforehand, until the line is noticed and given the same value as the phantasm. [7] In this way all the transitional stages between sensation and representation can be obtained. At no point do we come upon a psychical element that is absolutely incapable of being compared with the sensation, which we must undoubtedly regard as a physical object also. The way in which the presentations are connected by association is, however, quite different from the way in which the sensations are connected.

5.

Leonardo da Vinci discusses the mingling of phantasms with objects seen (see Ch. IV.) in the following words:

"I shall not omit to give a place among these directions to a newly-discovered sort of observation, which may, indeed, make a small and almost ludicrous appearance, but which is, nevertheless, very useful in awakening the mind to various discoveries. It consists in this, that thou shouldst regard various walls which are covered with all manner of spots, or stone of different composition. If thou hast any capacity for discovery, thou mayest behold there things which resemble various landscapes decked with mountains, rivers, cliffs, trees, large plains, hills and valleys of many a sort. Thou canst also behold all manner of battles, life-like positions of strange, unfamiliar figures, expressions of face, costumes, and numberless things which thou mayest put into good and perfect form. The experience with regard to walls and stone of this sort is similar to that of the ringing of bells, in the strokes of which thou willst find anew every name and every word that thou mayest imagine to thyself.

"Do not despise this opinion of mine when I counsel thee sometimes not to let it appear burdensome to thee to pause and look at the spots on walls, or the ashes in the fire, or the clouds, or mud, or other such places; thou wilt make very wonderful discoveries in them, if thou observest them rightly. For the mind of the painter is stimulated by them to many new discoveries, be it in the composition of battles, of animals and human beings, or in various compositions of landscapes, and of monstrous things, as devils and the like, which are calculated to bring thee honor. For through confused and undefined things the mind is awakened to new discoveries. But take heed, first, that thou understandest how to shape well all the members of the things that thou wishest to represent, for instance, the limbs of living beings, as also the parts of a landscape, namely the stones, trees, and the like."

All marked and independent appearance of phantasms without excitation of the retina - dreams and the half-waking state excepted - must, by reason of their biological purposelessness, be accounted pathological. In like manner, we are constrained to regard every abnormal dependence of phantasms upon the will as pathological.

Such, very likely, are the states that occur in insane persons who regard themselves as very powerful, as God, etc. But the delusions of the megalomaniac can equally be produced by the mere absence of inhibitory associations; for instance, one can believe in a dream that one has solved the most tremendous problems, because the associations that reveal the contradiction do not take place.

6.

After these introductory remarks we may now turn to the consideration of a few physiologico-optical phenomena, the full explanation of which, it is true, is still distant, but which are best understood as the expressions of an independent life on the part of the sense-organs.

We usually see with both eyes, and agreeably to definite needs of life, not colors and forms, but bodies in space. It is not the elements of the complex, but the whole physiologico-optical complex that is of importance. This complex the eye seeks to fill out and supplement, according to the habits acquired (or inherited) in its environment, whenever, as a result of special circumstances, the appearance of the complex is incomplete. This occurs oftenest in monocular vision, but is also possible in the binocular observation of very distant objects where the stereoscopic differences consequent upon the distance of the eyes from each other vanish.

We generally perceive, not light and shadow, but objects in space. The shading of bodies is scarcely noticed. Differences in brightness produce differences in the sensation of depth, and help to produce the modelling of bodies when the stereoscopic differences are insufficient for this purpose, a condition which is very noticeable in the observation of distant mountains.

Very instructive, from this point of view, is the image on the dull plate of the photographic camera. We are often astounded at the brightness of the lights and the depth of the shadows, which were not noticed in the bodies themselves as long as one was not compelled to see everything in a single plane. I remember quite well that, in my childhood, all shading of a drawing appeared to me an unjustifiable disfigurement, and that an outline-drawing was much more satisfactory to me. It is likewise well known that whole peoples, for instance the Chinese, despite a well-developed artistic technique, do not shade at all, or shade only in a defective manner.

The following experiment, which I made many years ago, [8] illustrates very clearly the relation in question between the sensation of light and the sensation of depth. We place a visiting-card, bent crosswise before us on the desk, so that its bent edge $b\,e$ is towards us. Let the light fall from the left. The half $a\,b\,d\,e$ is then much lighter, the half $b\,c\,e\,f$ much darker - a fact which is, however, scarcely perceived in unprejudiced observation. We now close one eye. Hereupon, part of the space-sensations disappear. Still we see the bent card spatially and nothing noticeable in the illumination. But as soon as we succeed in seeing the bent edge depressed instead of raised, the light and the shade stand out as if painted thereon. I pass over for the

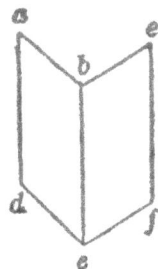

Fig. 24.

moment the perspectival reversal of the card, which can easily be explained. Such an "inversion" is possible, because depth is not determined by a monocular image. If in Fig. 25, 1 O represents the eye, $a\,b\,c$ a section of a bent card, and the arrow the direction of the light, $a\,b$ will appear lighter than be.

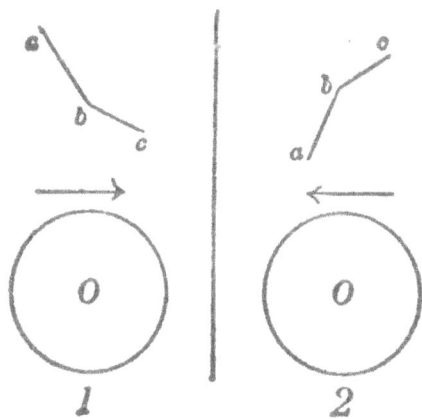

Fig. 25.

Also in 2, $a\,b$ will appear lighter than $b\,c$. Plainly, the eye must acquire the habit of varying the fall in the sensation of depth concomitantly with the change in brightness of the surface-elements that it sees. The fall and the depth diminish, with diminishing illumination, towards the right, when the light falls from the left (1); contrariwise, when it falls from the right. Since the wrappings of the bulb in which the retina is embedded are translucent, it is not a matter of indifference for the distribution of light

123

upon the retinae whether the light falls from the right or the left. Accordingly, things are so arranged that, without any aid of the judgment, a fixed habit of the eye is developed, by means of which illumination and depth are connected in a definite way. If now, by virtue of another habit, it is possible to bring a part of the retina into conflict with the first habit, as in the above experiment, the effect is made manifest in remarkable sensations. Certain experiments of Fechner's have shown how important the effect of the light that penetrates through the wrappings of the bulb can become. [9] One observation in this connexion is as follows. Beneath my writing-table is a grey-green rug, a small piece of which I can see as I write. Now, when a double image of this bit arises, accidentally or intentionally, when the sunlight or daylight comes from the left, the image belonging to the left, or more strongly illuminated eye, is a vivid green by contrast, while the image on the right side is quite dull in color. It would be interesting to study the variation of intensity and color of the illumination of the bulb in the case of these images and in experiments in inversion.

The purpose of the preceding remarks is merely to point out the character of the phenomenon under consideration and to indicate the direction in which a physiological explanation (exclusive of psychological speculation) is to be sought. We will further remark that, with respect to interchangeable qualities of sensation, a principle similar to that of the conservation of energy seems to hold. Differences of brightness are partly transformed into differences of depth, and themselves become weaker in the process. At the expense of differences of depth, on the other hand, differences of brightness may be augmented. An analogous observation will be made later on in another connexion.

7.

The habit of observing bodies as such, that is, of giving attention to a large and spatially cohering mass of light-sensations, is the cause of peculiar and often surprising phenomena. A two-colored painting or drawing, for instance, appears in general quite different according as we take the one or the other color as the background. The puzzle pictures, in which, for example, an apparition makes its appearance between treetrunks as soon as the dark trees

Fig. 26.

are taken as the background, and the bright sky as the object, are well known. In exceptional instances only do background and object possess the same form a configuration frequently employed in ornamental designs, as may be seen in Fig. 26, taken from p. 15 of the above mentioned *Grammar of Ornament,* also in Figs. 20 and 22 of Plate 45, and in Fig. 13 of Plate 43 of that work.

8.

The phenomena of space-vision which accompany the monocular observation of a perspective drawing, or, what amounts to the same thing, the monocular observation of an object, are generally very lightly passed over, as being self-evident in nature. But I am of the opinion that there is yet much to be investigated in these phenomena. One and the same image in perspective may represent an unlimited number of different objects, and consequently the space-sensation can be only in part determined by such a drawing. If, therefore, despite the many bodies conceivable as belonging to the figure, only a few are really seen with the full character of objectivity, there must exist some good physiological reason for the fact. It cannot arise from the adducing of auxiliary considerations in thought, nor from the awakening of conscious remembrances in any form, but must depend on certain organic life habits of the visual sense. If the visual sense acts in conformity with the habits which it has acquired under the conditions of life of the species and the individual, we may, in the first place, assume that it proceeds according to the principle of probability; that is, those functions which have been most frequently excited together before, will afterwards tend to make their appearance together when only one is excited. For example, those particular sensations of depth which in the past have been most frequently associated with a given perspective figure, will be readily reproduced again when that figure makes its appearance, although not necessarily co-determined thereby. Furthermore, a principle of economy appears to manifest itself in the observation of perspective drawings; that is to say, the visual sense never of itself burdens itself with any greater effort than is demanded by the stimulus. The two principles coincide in their effects, as we shall presently see.

9.

The following may serve as a detailed illustration of the above. When we look at a straight line in a perspective drawing, we always see it as a straight line in space, although the straight line, qua perspective drawing, may correspond to an unlimited number of different plane curves, qua objects. But only in the special case where the plane of a curve passes through the centre of one eye, will it be reproduced on the retina in question as a straight line (or as a great circle), and only in the yet more special case where the plane of the curve passes through the centres of both eyes, will it be reproduced as a straight line for both eyes. It is thus extremely improbable that a plane curve should ever appear a straight line, while on the other hand a straight line in space is always reproduced on both retinae as a straight line. The most probable object, therefore, answering to a straight line in perspective, is a straight line in space.

The straight line has various geometrical properties. But these geometrical properties, for example the familiar characteristic of being the shortest distance between two points, are not physiologically of importance. It is of far more consequence that straight lines lying in the median plane or perpendicular thereto are physiologically symmetrical to themselves. A vertical lying in the median plane is also physiologically distinguished by its perfect uniformity of depth-sensation, and by its coincidence with the direction of gravity. All vertical straight lines may be readily and quickly made to coincide with the median plane, and consequently partake of this physiological advantage. But the spatial straight line generally, must be physiologically distinguished by some further mark. Its sameness of direction in all its elements has already been pointed out. In addition to this, however, it is to be noted that every point of a straight line in space marks the mean of the depth-sensations of the neighboring points. Thus the straight line in space gives a minimum of departure from the mean of the depth-sensations, just as every point on the straight line gives the mean of the similar space-values of the adjacent points; and the assumption forthwith presents itself that the straight line is seen with the least effort. The visual sense acts therefore in conformity with the principle of economy, and, at the same time, in conformity with the principle of probability, when it exhibits a preference for straight lines.

As early as 1866, I wrote, in the *Proceedings of the Vienna Academy*, Vol. 54: "Since straight lines everywhere surround civilized human beings, we may, I think, assume, that every straight line which can possibly be produced upon the retina has been seen numberless times, in every possible way, spatially as a straight line. The efficiency of the eye in the interpretation of straight lines ought not, therefore, to astonish us." Even then I wrote this passage (opposing the Darwinian view, which I supported in the same paper) half-heartedly. To-day I am more than ever convinced that the efficiency referred to is not the result of individual practice, nor indeed of human practice at all, but that it is also characteristic of animals, and is, at least in part, a matter of inheritance.

10.

The deviation of a sensation from the mean of the adjacent sensations is always noticeable, and exacts a special effort on the part of the sense-organ. Every new turn of a curve, every projection or depression of a surface, involves a deviation of some space-sensation from the mean of the surrounding field on which the attention is directed. The plane is distinguished physiologically by the fact that this deviation from the mean is a minimum, or for each point in particular = o. In looking through a stereoscope at a spotted surface, the separate images of which have not yet been combined into a binocular image, we experience a peculiarly agreeable impression when the

whole is suddenly flattened out into a plane. The aesthetic impressions produced by the circle and the sphere seem to have their source mainly in the
fact that the abovementioned deviation from the mean is the same for all
points.

11.

That the deviation from the mean of the environment plays a role in light-
sensation I pointed out many years ago. [10] If a row of black and white sec-
tors, such as are shown in Fig. 27, be painted on a strip of paper $A A B B$ and
this be then wrapped about a cylinder the axis of which is parallel to AB,
there will be produced, on the rapid rotation of the cylinder, a grey field with
increasing illumination from B to A, in which, however, a brighter line $\alpha \alpha$,
and a darker line $\beta \beta$, make their appearance. The points which correspond
to the indentations a are not physically brighter than the neighboring parts,
but their light-intensity exceeds the mean intensity of the immediately adjacent parts, while, on the other hand, the light-intensity at β falls short of the
mean intensity of the adjacent parts. [11] This deviation from the mean is
thus distinctly felt, and accordingly imposes a special burden upon the organ
of sight. On the other hand a continuous change in brightness is scarcely noticed, as long as the brightness of each particular point corresponds to the
mean of the adjacent points. Long ago I drew attention to the important teleological bearing of this fact on the saliency and the delimitation of objects
(*Sitzungsberichte der Wiener Akademie* October 1865, and January 1868).
Small differences are slurred over by the retina, and larger differences stand
out with disproportionate clearness. The retina schematizes and caricatures.
At an even earlier period the important part which outlines play in vision
had been noticed by Panum.

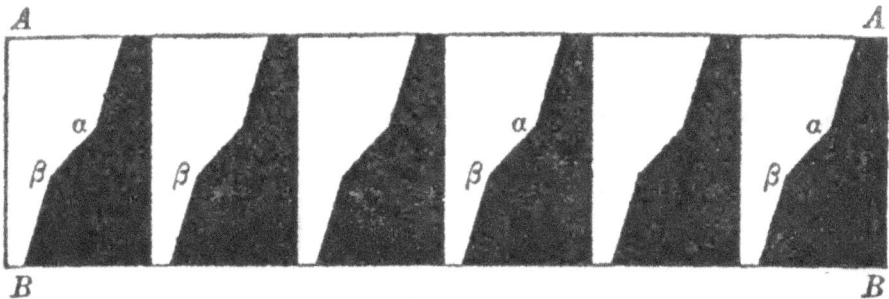

Fig. 27.

A series of very various experiments, of which that represented in Fig. 27
is one of the simplest, led me to the conclusion that the illumination of a position on the retina is felt in proportion to its deviation from the mean of the
illuminations of the adjacent positions. The value of the retinal positions in
determining this mean is to be conceived as rapidly decreasing with their
distance from the position under consideration, a fact which of course can

only be explained as depending on an organic reciprocal action of the retinal elements on one another. Let $i = f(x, y)$ be the intensity of illumination of the retina with reference to a system of co-ordinates (XY); then the mean value determining the intensity for a given position may be symbolically represented as approximately

$$i + \frac{m}{2}\left(\frac{d^2 i}{dx^2} + \frac{d^2 i}{dy^2}\right)$$

where m is constant, and the radii of all curves of the surface $f(x, y)$ are taken as large in proportion to the distance at which the retinal positions are still perceptibly influenced. Now according as $\left(\frac{d^2 i}{dx^2} + \frac{d^2 i}{dy^2}\right)$ positive or negative, the position on the retina experiences a darker or a brighter sensation respectively than it does under equal illumination of the adjacent positions with the intensity corresponding to itself. If the surface $f(x, y)$ has edges and indentations, $\left(\frac{d^2 i}{dx^2} + \frac{d^2 i}{dy^2}\right)$ becomes infinite, and the formula is useless. In this case, however, a marked increase of darkness or brightness corresponds to the indentation, though of course not an infinite increase or decrease. The increase or decrease, again, are not denned by a hard and fast line, but fade gradually away, as we should expect from the principle of deviation from the mean. For the retina consists, not of sensitive points, but of an infinite number of sensitive elements of finite extension. As regards the law of the reciprocal action of these elements, we still do not know it accurately enough to enable us to determine precisely the phenomena of this special case.

It is easy to go wrong in judging of the objective distribution of light according to the subjective impression, and consequently a knowledge of the above-mentioned law of contrast is important even for purely physical researches. Thus Grimaldi was deceived by a phenomenon of this kind. We come across the same phenomenon in the investigation of shadows, and of spectral absorption, and in countless other cases. Peculiar circumstances prevented my papers on this subject from becoming generally known, and the relevant facts were discovered for the second time thirty years later. [12]

It may seem surprising that, in addition to i, the second differential quotients of i, but not the first, $\frac{di}{dx}$, $\frac{di}{dy}$, seem to influence the sensation of brightness. We scarcely notice a regular and continuous rise in the intensity of illumination of a surface, for instance, in the direction x, - and special devices are necessary to convince one that there is a rise. On the other hand, these first differential quotients exercise an influence on the modelling, on the plastic quality, of the surface seen. Call the horizontal direction x, and

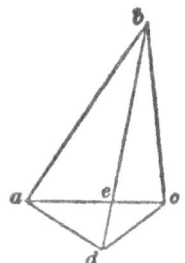

Fig. 28.

the distance as regards depth of a point on the illuminated surface τ; then $\overline{\frac{di}{dx}}$ and $\overline{\frac{d^2\tau}{dx^2}}$ are parallel. This expression, which of course is only to be understood symbolically, means that we have the representation of a cylindrical surface with vertical generatrix and plane horizontal directrix $\tau = F(x)$, of which the second differential quotients $\overline{\frac{d^2\tau}{dx^2}}$ (curvatures) are parallel to the first differential quotients, - the rises in intensity of illumination. The tracing of the curve is determined by the accessory circumstances indicated earlier this chapter.

12.

With regard to the depth-sensations excited by a monocular image, the following experiments are instructive. Fig. 28 is a plane quadrilateral with its two diagonals. If we regard it monocularly, it is most easily seen, according to the laws of probability and economy, as a plane. In the great majority of cases, objects which are not plane, force the eye to the vision of depth. Where this compulsion is lacking, the plane object is the most probable and at the same time the most convenient for the organ of sight.

The same drawing may be also viewed monocularly as a tetrahedron, the edge b d of which lies in front of a c, or as a tetrahedron, the edge b d of which lies behind a c. The influence of the imagination and the will upon the visual process is extremely limited; it is restricted to the directing of the attention and to the selection of the appropriate disposition of the organ of sight for one of a number of cases given by habit, of which, however, each one, when chosen, takes its place with mechanical certainty and precision. Looking at the point e, we can, as a fact, produce either of the two optically possible tetrahedrons at will, according as we represent to ourselves b d as nearer or farther away than a c. The organ of sight is practised in the representation of these two cases, since it often happens that one body is partly covered by another.

Loeb [13] thinks that the act of bringing Fig. 31 nearer to the eyes gives rise to short-distance accommodation, and thereby also to our seeing the fixed edge b e as raised. I have not been able to obtain any such definite result myself, nor can I find any sufficient theoretical ground for it, [14] although I readily admit that changes in the distance of the figure easily lead to changes in our view of it.

The same figure may, finally, be seen as a four-sided pyramid, if we imagine the conspicuously situated point of intersection e before or behind the plane a b c d. This is difficult to do, if b e d and a e c are two perfectly straight lines, because it conflicts with the habit of the organ of sight to see, without constraint, a straight line bent; the effort is successful only because the point e has a conspicuous position. But if there is a slight indentation at *e*, the attempt involves no difficulty.

The effect of a linear perspective drawing is felt as unerringly by one who is ignorant of perspective as by one who is thoroughly conversant with the theory, provided he is able to disregard the plane of the drawing, a condition readily fulfilled in monocular observation. Reflexion, and even the remembrance of seen objects, have, according to my belief, little or nothing to do with the effect in question. Why the straight lines of a drawing are seen as spatial straight lines, has already been pointed out. Where straight lines appear to converge to a point in the plane of the drawing, the converging or approaching ends are transferred, according to the principles of probability and economy, to like or to nearly like depth. This gives us the effect of vanishing points. It is possible to see such lines as parallel, but there is no necessity for such an impression. If we hold the drawing, Fig. 29, on a level with the eye, it may represent to us a

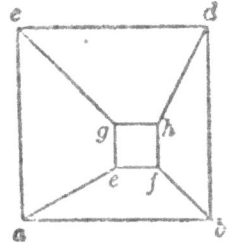

Fig. 29.

glance down a passage-way. The ends *g h e f* are transferred to like distances. If the distance is great, the lines *a e, b f, c g, d h* appear horizontal. If we raise the drawing, the ends *e f g h* rise, and the floor *a b e f* seems to have an upward slope. Upon lowering the drawing, the opposite phenomenon is presented; and analogous changes may be observed by moving the drawing towards the right or the left. In these facts, the elements of perspective effect find simple and clear expression.

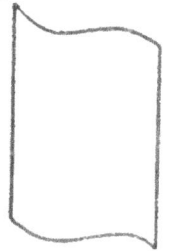

Fig. 30.

Plane drawings, provided they consist entirely of straight lines, everywhere intersecting each other at right angles, almost always appear plane. If oblique intersections and curved lines occur, the lines easily pass out of the plane; as is shown, for example, by Fig. 30, which may, without difficulty, be conceived as a curved sheet of paper. When outlines, such as are represented in Fig. 30, have assumed definite spatial form, and are seen as the Fig 30. boundary of a surface, the latter, to describe it briefly, appears as flat as possible, that is to say, is presented with a minimum of deviation from the mean of the depth-sensation. [15]

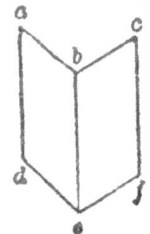

Fig. 31.

13.

The peculiar reciprocal action of lines intersecting obliquely in the

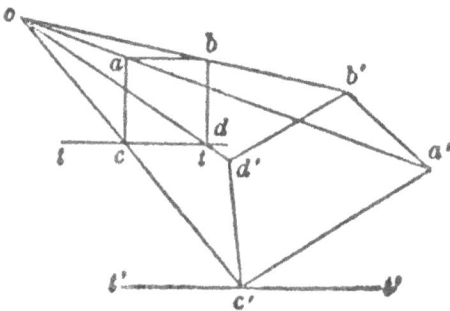

Fig. 32.

plane of the drawing (or on the retina), whereby such lines are mutually forced out of the plane of the drawing (or out of the plane perpendicular to the line of sight) was first observed by me on the occasion of the above-mentioned experiment with the monocular inversion of a card. The card in Fig. 31, whose edge *b e* when turned outwards towards me is in a vertical position, assumes, when I succeed in seeing *b e* depressed, a recumbent position, like that of a book lying open upon my table, with the result that *b* appears further away than *e*. When one is once acquainted with this phenomenon, the inversion may be performed with almost every object, and one can always observe along with the change of form or tilting over, this remarkable simultaneous change of position. The effect is especially astonishing in the case of transparent objects. Let abed be a section of a glass cube lying on a table / /, and let O be the eye (Fig. 32). On monocular inversion, the angle a is projected to a 1 ', b to the nearer point V ', t to </, and *d* to *d'*. The cube will seem to stand obliquely on its edge c f upon the table t' t'. In order that the drawing might afford a better survey of the phenomenon, the two images have been represented behind, not within one another. If a drinking-glass partly filled with a colored liquid be substituted for the cube, it will be seen, together with the surface of the liquid, in a similar oblique position.

With sufficient attention, the same phenomena may be observed with any linear drawing. If we place the page containing Fig. 31 vertically before us, and observe it monocularly, we shall see *b* project if *b e* be raised, but if *b e* be depressed b will retreat and e will project and come nearer to the observer. Loeb [16] notices that when this happens the points *a e* remain in the plane of the drawing. And, in fact, this makes the change of direction intelligible. If we draw the dotted lines, as in Fig. 32A, and imagine the figure, so far as it lies outside the dotted triangle, obliterated, we are

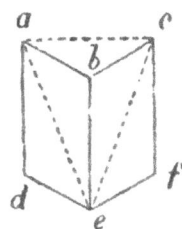

Fig. 32 a

left with the image of a hollow or raised three-sided pyramid, which lies with its base in the plane of the drawing. Inversion no longer produces any sort of mysterious change of position. It would seem, therefore, that every point seen monocularly aims at the minimum deviation from the mean of the sensation of depth, which is attainable under the conditions of the experiment, and that the whole object seen aims at the minimum attainable amount of removal from Hering's nucleus-surface.

When we consider the deformations which a plane rectilinear figure undergoes when traced in monocular space, all such deformations may be qualitatively reduced to the following principle: the legs of an acute angle are thrust out on opposite sides of the plane of the drawing, or of the plane perpendicular to the line of sight, and the legs of an obtuse angle are thrust out on the same side. In this process acute angles are magnified and obtuse angles diminished. All angles tend to become right angles.

14.

This principle suggests that the phenomenon just described is closely related to Zöllner's pseudoscopy and the numerous phenomena connected with it. Here again everything turns on the apparent enlargement of acute and the apparent reduction of obtuse angles, except that the drawings are seen in the plane. But when they are seen in monocular space the pseudoscopic effects vanish, and the phenomena described above appear. Now although these phenomena have been much studied, no completely satisfactory explanation of them has as yet been offered. Naturally such superficial explanations as, for instance, the assumption that we are chiefly accustomed to see right angles, are inadmissible, if the investigation is not utterly to miscarry or to be prematurely broken off. We see oblique-angled objects often enough, but never, without artificial preparation, the surface of a liquid at rest and yet oblique, as we did in the experiment given above. Yet, the eye, it would seem, prefers the oblique liquid surface to an oblique-angled body.

The elemental power displayed in these processes has, I believe, its root in far simpler habits of the organ of sight, - habits whose origin doubtless antedates the civilized life of man. I once tried to explain the phenomena in question by a contrast of directions analogous to the contrast of colors, but without arriving at a satisfactory result. But more recent researches by Loeb, [17] Heymans, [18] and others, and observations by Hoefler [19] on curve-contrasts, are very much in favor of a theory of contrast. Moreover, quite lately at any rate, there is a decidedly increased tendency to adopt some purely physiological explanation. [20]

The principle of economy, again, has afforded me no enlightenment as far as Zöllner's pseudoscopy is concerned.

A somewhat greater prospect of success seems to be offered by the principle of probability. Let us conceive the retina as a perfect sphere and imagine the eye fixed upon the vertex of an angle a in space. The planes passing through the centre of the eye and the lines containing the angle, project these lines upon the retina, describing thereon a spherical segment having the angle A, which represents the angle of the monocular image. Now an infinite number of values for a, varying from o to 180, may correspond to a constant value of A, as will be seen if we reflect that the lines including the objective angle may assume every possible position in the planes of their projection. Consequently, to a seen angle A, we may have corresponding all the possible values of the objective angle a that can be obtained by causing each of the sides, b and c, of the triangle to vary between 0° and 180°. The actual result is, supposing the calculation to be performed in a definite manner, that larger angles are the most probable objects corresponding to observed acute angles, and smaller angles the most likely counterparts of observed obtuse angles. I was not, however, in a posi-

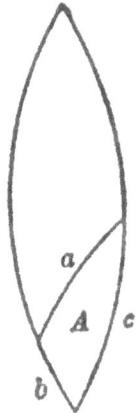

Fig. 33.

tion to determine whether those cases, which we are inclined to regard as geometrically equally probable, - ought also to be regarded as physiologically equally probable - a question which is both essential and important. Moreover, the whole conception has a much too artificial cast for me.

15.

I cannot refrain from mentioning here the attempt which has been made by A. Stohr to reach an explanation of the phenomena described above from an entirely new point of view. With the general considerations by which he was guided I am in full sympathy and agreement. On the other hand, I have not yet been able to convince myself that there is a demonstrable foundation of fact corresponding to his hypotheses. Moreover, the relations which he presupposes are so complicated that it is not easy to decide this question, without thoroughly covering the ground experimentally oneself. I therefore do not know whether Stöhr's views will amount to a complete explanation on all points. In one of his less recent works [21] the assumption is made that to the dioptric image of the eye in front of the retina, there corresponds a catoptric image in the retina, the latter having relief in proportion to the depth of the former. Depth in the retina is thus both the determining factor for the sensation of depth in visual space, and the regulating factor in accommodation. As a matter of fact, I have always asked myself what the means could be by which the direction of change of accommodation is determined; for change of accommodation cannot be determined merely by the magnitude of the circle of dispersion; also there is only a loose connection between accommodation and convergence, and moreover a single eye by itself is accommodated. Against this view, on the other hand, must be set the numerous observations which have been made as to the worthlessness of accommodation for the sensation of depth. The great thickness of the retina in the eyes of insects [22] suggests, again, that the perception of relief may be connected with some function of the retina.

In two later works [23] he goes further, taking this theory as a basis. In the second of these books we find a view not unlike Scheffler's, but in a more physiological form. The dominant view, according to which the images of positions which deviate more or less from the corresponding positions are fused into one unified impression, Stohr thinks is untenable. "Where is the pointsman who arranges the change in such a way that, not only in extraordinary cases, but also as of set purpose, two stimuli may be brought into combination in the central organ over a quite unfamiliar pair of lines of conduction?" It is assumed that the retinae of both eyes naturally endeavour to minimalize the light-stimulus, and thus tend towards the equalization of unequal images. The nervous elements excite the ciliary muscle, doing this not only in a quite regular and uniform manner, but also, according to requirements, with great irregularity. Regular contraction of the ciliary muscle produces a greater bulging of the lens and slight contraction of the retina. If in this process the retinal elements carry their position-values with them, the same retinal image appears enlarged. In this way, according to Stohr, we can understand why Panum's proportional systems of circles (up to circles with radii in the proportion of 4 to 5) are seen, in virtue of the mutual adaptation of the two eyes, with identical parts of the retina, as simple and as having a size which is the mean of their sizes. By depicting one system in red and the other in green points alternately, so that in the united binocular image the red points appear between the green, Stöhr proves that the fusion of the systems of circles is not caused by the suppression of one of the images. And irregular contraction of the ciliary muscle is supposed to produce various effects, - (1) an irregular deformation of the lens with very various displacements of the apices of the diacaustic of different pencils of rays, whereby change in the relief of the dioptric and catoptric images is produced; and (2) various minimal deformations of the retina. Stohr thinks that he can demonstrate the possibility of his theory by detailed calculations, and that he can prove the actuality of his presuppositions by investigating subjects with eyes in which the crystalline lens is absent or out of its proper position (aphakia). In any case his theory has led to experiments with surprising results, - for instance, the stereoscopic indentation of straight lines, - and, if only on that account, it deserves to be considered with respect. But although his whole conception of the eye and its parts as living organisms is extremely congenial to me, I have not yet been able to convince myself that the assumptions which he makes in order to explain more complicated cases of spatial vision everywhere fulfil their purpose. [24]

Stohr's departure from the traditions of physiological optics is very great. In itself this can be no reason for refusing to test his theory closely, especially since S. Exner's and Theodor Beer's [25] researches in comparative physiology, which have been so rich in beautiful and remarkable results, have made us familiar with eyes characterized by a complexity and variety of organic adaptation such as a physicist would scarcely have supposed possible *a priori*. It is possible that Stohr's views may apply to other organs of vision, although perhaps not to the human eye.

There are many phenomena that make it probable that the act of sight involves other processes of change in the eye which still require to be investigated. Stereoscopic images with prominent stereoscopic differences display, when they are gazed at for a long time, an enormous increase in the growth of their relief by successive stages, even though fusion has apparently been complete for some time. Wave-like curvatures and swellings have been observed in systems of fine, smooth, parallel lines, and these have been explained in a rather peculiar manner as referable to the incapacity of the mosaic-like texture of the retina to reproduce straight lines of such fineness. I have, however, always noticed this phenomenon when I have gazed for some time at systems of straight lines which are clearly visible and by no means micrometric. Thus the mosaic of the retina can have nothing to do with the matter. I should prefer to suppose that the exertion involved, perhaps by means of small displacements in Stohr's sense, introduces a certain disorder into the space-values. [26]

16.

The ease of the transition from the process of seeing plane figures pseudo-scopically to that of seeing them monocularly in space will probably help us to throw light upon the former. This conjecture is confirmed by the following facts. A plane linear drawing, monocularly observed, usually appears plane. But if the angles be made to vary and motion be introduced, every drawing of this sort will assume a solid form. We then generally see a solid body in rotation, such as I have described on a former occasion. [27] The well-known vibrating acoustic figures of Lissajous, which on varying their difference of phase, appear to lie on a revolving cylinder, afford a beautiful example of the process in question.

Here, again, reference might be made to our habit of constantly dealing with solid bodies. In fact, solid bodies engaged in revolutions and turnings continually surround us. Indeed, the whole material world in which we move is, to a certain extent, a single solid body; and without the help of solid bodies we could never attain to the conception of geometrical space. We do not generally notice the position of the single points of a body in space, but apprehend its dimensions directly. Herein lies, for the unpractised, the main difficulty of drawing a perspective picture. Children, who are accustomed to *seeing* bodies in their real dimensions, do not understand perspective fore-shortenings, and are far better satisfied with simple outlines or silhouettes. I can well remember this condition of mind, and through this remembrance am able to comprehend the drawings of the ancient Egyptians, which represent all parts of the body as far as possible in their true dimensions, thus pressing them, as it were, into the plane of the drawing, as plants are pressed in a herbarium. In the Pompeian wall-paintings, too, we still meet with a perceptible dislike for foreshortening, although here the sense of perspective is already manifest. The old Italian masters, on the other hand, in the consciousness of their perfect mastery of the subject, often amuse themselves with excessive and sometimes even unbeautiful foreshortenings, which occasionally demand considerable exertion of the eye.

17.

There can be no question, therefore, but that we are much more familiar with the process of seeing solid bodies with the distances between their salient points unchanged, than with the process of separating out their depth, which is always the result, in the first place, of deliberate analysis. Accordingly, we may expect that wherever a coherent mass of sensations, which, in virtue of its continuous transitions and its common coloring, merged into a unity, exhibits spatial alteration, the change will be seen preferably as the motion of a solid body. I must confess, however, that this way of looking at the matter does not satisfy me. I believe, rather, that here, too, an elementary habit of the organ of sight is at the root of the matter, - a habit which did not originally arise through the conscious experience of the individual, but, on the contrary, antecedently facilitated our apprehension of the movements of solid bodies. If we should assume, for example, that every diminution of the transverse dimension of an optical sensation-mass to which the attention was directed had the tendency to induce a corresponding augmentation of the dimension of depth, and *vice versa*, we should have a process quite analogous to that which we have already considered above and which was compared with the conservation of energy. This view is certainly much simpler and supplies an equally adequate explanation. Furthermore, it enables us to comprehend more easily how such an elementary habit could be acquired, how it could find expression in the organism, and how the disposition towards it could be inherited.

As a sort of counterpart to the rotation of solid bodies exhibited to us by the organ of sight, I will cite here an additional observation. If an egg, or ellipsoid with dull, uniform surface be rolled over the top of a table, but in such manner that it does not turn about its axis of generation, but performs jolting movements, we shall fancy we see, on viewing it binocularly, a liquid body, or large oscillating drop. The phenomenon is still more noticeable if the egg, with its longitudinal axis in a horizontal position, be set in moderately rapid rotation about a vertical axis. This effect is immediately destroyed when marks, whose movements we may follow, are made upon the surface of the egg. A rotating solid body is then seen.

The explanations offered in this chapter are certainly far from complete, yet I believe that the considerations adduced will have some effect in stimulating and preparing the way for a more exact and thorough study of these phenomena.

[1] A. Bethe. "Dürfen wir den Ameisen und Bienen psychische Qualitäten zuschreiben?" Pflüger's *Archiv,* Vol. LXX., p. 17; "Noch einmal über die psychischen Qualitaten der Ameisen," *Ibid.,* Vol. LXXIX., p. 39; Beer, Be the and Uexkiill, "Vorschlage zu einer objektivierenden Nomenklatur in der Physiologic des Nervensystems," *Centralblatt für Physiologie,* 1899, Vol. XIII., No. 6; H. E. Hering, "Inwiefern ist es möglich die Physiologie von der Psychologic sprachlich zu trennen?" *Deutsche Arbeit,* 1st year, No. 12.
[2] E. Wasmann, *Die psychischen Fähigkeiten der Ameisen,* Stuttgart 1899 (Zoologica, No. 26); *Vergleichende Studien über das Seelenleben der Ameisen ttnd der höheren Tiere,* 2nd edition, Freiburg im Breisgau, 1900.
[3] H. V. Buttel-Reepen, *Sind die Bienen Reflexmaschinen?,* Leipzig, 1900.
[4] A. Forel, "Psychische Fähigkeiten der Ameisen," *Transactions of the Fifth International Zoological Congress,* Jena, 1902; "Expériences et remarques critiques sur les sensations des insectes," *Rivista di Scienze Biologiche,* Como, 1900-1.

[5] Mach, *Populär-wissenchaftliche Vorlesungen*, "Ueber den Einfluss zufälliger Umstände," etc. , Leipzig, 1903, 3rd edition, pp. 294-295; *Prinzipien der Wärmelehre*, Leipzig, 1900. See the chapter on "Sprache und Begriffe."

[6] Cf. the fourth edition of this work, p. 153.

[7] Scripture, *The New Psychology*, London, 1897, p. 484.

[8] "Ueber die physikalische Wirkung räumlich verteilter Lichtreize," *Sitzungsberichte der Wiener Akademie, Vol. LIV.*, October 1866.

[9] Fechner, "Ueber den seitlichen Fenster- und Kerzenversuch," *Berichte der Leipziger Gesellschaft der Wissenschaften*, 1860.

[10] "Ueber die Wirkung der räumlichen Vertheilung des Lichtreizes auf die Netzhaut," *Sitzungsberichle der Wiener Akademie* (1865), Vol. LII. Continuation of the same inquiry: Sitzber. (1866), Vol. LIV.; Sitzber. (1868), Vol. LVII.; *Vierteljahrsschrift für Psychiatrie*, Neuwied-Leipzig, 1868 ("Ueber die Abhangigkeit der Netzhautstellen von einander").

[11] A remark concerning the analogies between light-sensation and the potential function will be found in my note "Ueber Herrn Guébhards Darstellung der Aequipotentialkurven," *Wiedemann's Annalen*, 1882, Vol. XVII., p. 864; and see my *Prinzipien der Wärmelehre*, 2nd edition, 1900, p. 118.

[12] H. Seeliger, "Die scheinbare Vergrosserung des Erdschattens bei Mondfinsternissen," *Abhandlungen der Münchener Akademie*, 1896; H. Haga and C. H. Wind, "Beugung der Röntgenstrahlen," Wiedemann's *Annalen*, Vol. LXVIII., 1899, p. 866; C. H. Wind, "Zur Demonstration einer von E. Mach entdeckten optischen Täuschung," Riecke and Simon's *Physikalische Zeitschrift*, I., No. 10. A. von Obermayer ("Ueber die Säume um die Bilder dunkler Gegenstände auf hellem Hintergrunde," Eder's *Jahrbuch der Photographie*, 1900) published a number of new facts which can be explained on the law of contrast laid down in the text. But, of my four papers, he is only acquainted with the first, and consequently states the law in its earlier defective form.

[13] Loeb, "Ueber optische Inversion," Pflüger's *Archiv*, Vol. XL., 1887, p. 247.

[14] Hillebrand ("Verhältnis von Akkommodation und Konvergenz zur Tiefenlokalisation," *Zeitschrift für Psychologie und Physiologie der Sinnesorgane*, Vol. VII., p, 97) has proved the slight importance which accommodation has for the seeing of depth.

[15] Here again, the depth-sensation resembles the potential function, in a space at the boundaries of which it is determined. This flat-aspossible surface does not coincide with the surface of minimal area, which would be obtained if the spatial outlines were made of wire, and then dipped in soap-suds, producing a Plateau's liquid film.

[16] Loeb, "Ueber optische Inversion," quoted earlier this chapter.

[17] Loeb, Pflüger's *Archiv*, 1895, p. 509.

[18] Heyman's *Zeitschrift für Psychologie und Physiologie der Sinnesorgane*, Vol. XIV., p. 101.

[19] Hoefler, *Ibid.*, Vol. XII., p. 1.

[20] Witasek, *Ibid.*, Vol. XIX,, p. 1.

[21] *Zur nativistischen Behandlung des Tiefensehens*, Vienna, 1892.

[22] Exner, *Die physiologie der facettierten Augen*, p. 188, Vienna, 1891.

[23] *Zur Erklärung der Zöllnerschen Pseudoskopie,* Vienna, 1898; *Binokulare Figurmischimg und Pseudoskopie,* Vienna, 1900.

[24] The following book has subsequently appeared: A. Stohr, *Grundfragen der psycho-physiologischen Optik,* Leipzig and Vienna, 1904. The problems in question are here discussed further.

[25] Th. Beer, "Die Akkommodation des Fischauges," Pflüger's *Archiv,* Vol. LVIII., p. 523; "Akkommodation des Auges in der Tierreihe," *Wiener klinische Wochenschrift,* 1808, No. XLII.; "Ueber primitive Sehorgane," Ibid., 1901, Nos. XL, XII., XIII.

[26] "Ueber die physiologische Wirkung räumlich verteilter Lichtreize," *Wiener Sitzungsberichte,* 2nd part, October 1866, pp. 7, 10, of the offprint.

[27] "Beobachtungen über monokulare Stereoskopie," *Sitzungsberichte der Wiener Akademie* (1868), Vol. LVIII.

XI. Sensation, Memory and Association

1.

THE foregoing discussions have shown beyond all possible doubt that out of mere sensations no psychical life resembling ours even in the remotest degree could be constituted. When a sensation is forgotten the moment after it has vanished, the only possible result is a disconnected mosaic and series of psychic states, such as we have to suppose in the case of the lowest animals and the most degraded idiots. At this stage, a sensation which does not have some such effect as to stimulate violently to movement - a sensation of pain, for instance - will scarcely receive attention. For instance, the sight of a vividly-colored spherical body, which is not supplemented by a memory of smell and taste, by memory, in a word, of the properties of a fruit and the experiences connected with a fruit, - remains unintelligible and is devoid of interest, in the manner that has been observed in "psychic blindness." The storing up and connexion of memories, and their power to evoke one another, - in short, Memory and Association, - are the fundamental requirements of a developed psychical life.

What is memory? A psychical event leaves psychical traces behind it, but it also leaves physical traces. Physically, as well as psychically, a child that has been burnt, or stung by a wasp, behaves in quite a different way from a child that has not had this experience. For the psychical and the physical are different only according to the way in which they are regarded. Nevertheless it is extremely difficult to discover in the physical phenomena of the inorganic world characteristics having any affinity to memory.

In the physics of the inorganic world everything seems to be determined by the circumstances of the moment, and the past seems to be entirely without any influence. The oscillations of a pendulum are equal, whether it is performing its first oscillation or whether 1000 others have already taken place. Hydrogen combines with chlorine in the same way, no matter whether it was previously in combination with bromine or with iodine. There are indeed, even in the physical sphere, cases in which the influence of the past is clearly expressed. The earth reveals to us the history of its geological past, and the moon does the same. My friend E. Suess has shown me a piece of rock marked with a system of very peculiar congruent parallel fissures, which he very plausibly interprets as a prehistoric seismogram.

2.

A wire notices, so to speak, during a considerable time, every torsion that it sustains. Every spark of a discharge is an individual, and is influenced by the discharges that have preceded it. The insulating layer of the Leyden jar preserves a history of all the previous charges.

The apparent contradiction is solved when we remember how in physics we are accustomed to idealize and schematize in an extreme degree the cases under consideration, always presupposing the simplest possible circumstances. If we assume a mathematical pendulum, then no doubt the thousandth oscillation is as the first, and no traces of the past are visible, precisely because we disregard such traces. But a real pendulum wears away its knife-edge, and is heated by internal and external friction, so that no oscillation, accurately considered, is exactly like any other. The result of every second and third torsion of a wire is somewhat different from what it would have been if there had been no previous torsions. If a similar schematization were possible in psychology, we should have men who behave identically and do not betray any of the influence of individual experiences.

In reality every psychical process leaves indelible traces behind, just as every physical process does. In both spheres there are irreversible processes: entropy increases, or the bond of a friendship that has been broken, and then renewed, is felt. And every real process contains at any rate some irreversible components.

3.

Now it will be said, and with justice, that traces of the past are still far from being the same thing as memory. As a matter of fact, what is required to increase the resemblance is, that processes which have taken place in the past should be set up afresh by some slight impulse. Old violins that have been well played on, Moser's electrical images (that come out when breathed upon), and the phonograph, afford rather better examples. Still, violins and phonographs have to be played by external forces, while human beings and their memories play themselves. For organic beings are not rigid material systems; they are essentially forms of the dynamic equilibrium of currents of "matter" and "energy." Now it is the forms of the deviation of these currents from the state of dynamic equilibrium that always repeat themselves in the same way, according to the way in which they have once been introduced. Such variations in the forms of dynamic equilibrium have still been but little studied by inorganic physics. As a very rough example we may take changes in the flow of liquids, produced by some chance circumstance, and then maintained. If we screw up a tap so tightly that only a thin, quiet trickle of water comes but, a chance jolt is enough to disturb the unstable equilibrium of the trickle, and to cause the water to run out in drops with a persistent rhythm. Suppose that a chain, lying coiled up in a tub, is allowed to run over a roller acting as a sort of lever, and to fall into another tub at a lower level. If the chain is very long and the difference of level very great, the velocity may become considerable, and then the chain, as is well known, has the property, whenever it is made to take a loop, of keeping this loop suspended in the air for some time, and continuing its flow in this shape. All these examples are very inadequate analogies to the plasticity which organisms possess for the repetition of processes and series of processes.

The foregoing considerations are intended to show that a comprehension of memory on physical lines is not unattainable, although we are still very far removed from it. There can be no doubt that a considerable enlargement of the point of view of physical science by means of the study of organic beings is required, before physics is capable of such a task. The great richness of memory is founded, no doubt, upon the reciprocal interaction and connexion of the organs. Still, we probably must ascribe a rudimentary memory even to elementary organisms; and if so, the idea inevitably suggests itself that, every chemical process in the organ leaves traces behind which are favourable to the reappearance of the same process. [1]

4.

It is well known that a very prominent position is given, in psychology, to the laws of association. These laws can be reduced to a single law, which consists in saying that if two contents of consciousness, A and B, have once appeared simultaneously, one of them, when it arises, will evoke the other. And in fact it is much easier to understand physical life when we have recognized the constant recurrence of this fundamental feature. The differences of mental process, in simple memory of an experience, in serious occupation, and in the free exercise of fancy or day-dreaming, can easily be understood by means of the concomitant circumstances. [2] It would, however, be a complete mistake to try to reduce *all* (p. 201) psychical processes to associations acquired during the life of the individual. In none of its phases do we meet with the psyche as a *tabula rasa*. At the very least we should have to assume innate associations side by side with the acquired. The innate impulses, [3] which, to a psychology that is purely introspective and confined to itself, must necessarily appear to be innate associations, are reduced by biology to innate organic connexions, and, in particular, to nervous connexions. It is therefore worthwhile to inquire whether all associations, including those acquired by the individual, do not depend upon innate connexions, of which some have been strengthened by use. [4] But in any case we must also ask whether the processes for the connexion of which in highly differentiated organisms special paths have been evolved, are not rather primary facts that already exist in lower organisms, and whether it is not their repeated occurrence accompanied by one another that has led to the formation of the paths in question. [5] A rational psychology cannot, of course, be content with temporary associations; it will have to provide for fixed paths of connexion also. Again, room must be found for the possibility of spontaneous psychical processes, not due to association, which excite the neighboring parts of the nervous system, and, when they are of great violence, even spread over the whole nervous system. On the one hand hallucinations, on the other reflex movements, are examples from the sensational and motor spheres, to which there are probably corresponding analogies in other spheres.

5.

Theories of the reciprocal action of the parts of the central nervous system seem to be opposed to a view which has been expounded by Loeb, [6] partly on the basis of his own work, partly on that of Goltz and Ewald. This view deserves to be noticed. According to it, the tropisms of animals are not essentially different from those of plants, the only advantage secured by the nerves in the case of animals being the more rapid transference of stimulus. The life of the nervous system is reduced to segmental reflexes, the co-ordination of movements to reciprocal excitation and transference of stimulus, and the instincts to chains of reflexes. The snapping-reflex of the frog, for instance, sets free the swallowing-reflex. Organized centres of great complication are not assumed, but the brain itself is regarded as an arrangement of segments. At the bottom of all these theories there lies, so far as I can judge, a happily conceived and important effort to shake off the trammels of unnecessarily complicated assumptions impregnated with metaphysics. But I cannot agree with Loeb when he treats Darwin's phylogenetic research on the instincts as a fallacious and one-sided proceeding, which ought to be dropped and replaced by physico-chemical investigations. Research of that kind was, no doubt, not within Darwin's horizon. But it was precisely that fact which secured for him the freedom of vision necessary to his great and peculiar discoveries, which no physicist, qua physicist, could have made. We are, indeed, everywhere trying to obtain, where it is possible, some insight into the physical constitution of things, some acquaintance with their immediate, or causal, connexions. But it is far from being the case that this is already possible everywhere. And in cases where it is not possible, it would at all events only be another and a very dangerous piece of one-sidedness, to give up other fruitful points of view, which can always be regarded as provisional. The steam-engine can, as Loeb says, only be understood on physical lines. But this is only true of a particular given steam-engine. When it is a question of understanding the present *forms* of the steam-engine, physical considerations are not sufficient. The whole history of technical and social culture, and the geological presuppositions involved, must be taken into account. It is possible that, in the last resort, each one of these factors is susceptible of a physical explanation, but it has explained our difficulties long before that stage is reached. [7]

6.

If I can imagine that, while I am having sensations, I myself or someone else could observe my brain with all the necessary physical and chemical appliances, it would then be possible to ascertain with what processes of the organism sensations of a particular kind are connected. The question so often asked, what is the lower limit of sensation in the organic world, whether the lowest animals have sensations, or whether plants have, could then be brought nearer to its solution, at any rate so far as analogy goes. As long as this problem has not been solved in even one single special case, no decision of the question is possible. It is sometimes even asked whether inorganic "matter" has sensations. The question is natural enough, if we start from the commonly current physical conception which represents matter as the immediately and undoubtedly given reality out of which everything, inorganic and organic, is constructed; for sensation must either arise suddenly somewhere or other in this structure, or else have been present in the foundation-stones from the beginning. From our point of view the question is merely a perversion. Matter is for us not what is primarily given. What is primarily given is, rather, the elements, which, when standing to one another in a certain known relation, are called sensations. Every scientific problem that can have any meaning for a human individual is concerned with the ascertainment of the dependence of the elements on one another. What in every-day life we call matter is a definite kind of connexion between the elements. The question as to whether matter has sensations would therefore run as follows: does sensation belong to a definite kind of connexion between the elements, these elements themselves also being, when in a certain relation, always sensations? Put in this form; no one will want to ask the question. [8] Everything that can have any interest for us must be reached in the course of following out the general task of science. We ask whether animals have sensations, when the assumption of sensations helps us better to understand their behaviour as observed by means of our own senses. The behaviour of a crystal is already completely determined for our senses; and thus to ask whether a crystal has sensations, which would provide us with no further explanation of its behaviour, is a question without any practical or scientific meaning.

[1] Ostwald has made a bold attempt at a chemical theory of memory, based on his theories about katalysis. See his *Vorlesungen über Naturphilosophie,* 1902, pp. 369 sqq.

[2] Mach, *Erkenntnis und Irrtum,* 1905, pp. 29 sqq.

[3] The most striking of these, because they make their appearance at the moment when the mental faculties and the power of observation are fully developed, are the first manifestations of the sexual impulse. I have been told by a perfectly trustworthy man, a person with a strong love of truth, that when he was a

lad of sixteen, being quite innocent and inexperienced at the time, he saw a lady in a low-necked dress, and was startled to find that he was suddenly aware of a striking bodily change in his person; this change he took to be an illness, and consulted a colleague about it. The whole complex of entirely new sensations and feelings which were then suddenly revealed to him was colored by a strong additional element of fear.

[4] H. E. Ziegler, "Theoretisches zur Tierphysiologie und vergleichenden Neuro-physiologie," *Biologisches Zentralblatt,* Leipzig, 1900, Vol. XX., No. 1.

[5] If we think of organic life as a state of dynamic equilibrium of various chemical component-phases, in which, speaking generally, a disturbance of one component causes a disturbance of the rest, we shall then be justified in hoping to explain, not only memory, but association too, on chemical lines. See Note 1. and Chapter V.

[6] Loeb, *Vergleichende Physiologie des Gehirns,* Leipzig, 1899.

[7] Loeb, *Vergleichende Physiologie des Gehirns,* p. 130.

[8] Cf. Mach, *Populär-Wissenschaftliche Vorlesungen,* 3rd ed., 1903, p. 242.

XII. The Sensation of Time

[1]

1.

MUCH more difficult than the investigation of space-sensation is that of time-sensation. Many sensations make their appearance with, others without, a clear sensation of space. But time-sensation accompanies every other sensation, and can be wholly separated from none. We are referred, therefore, in our investigations here, to the variations of time-sensation. With this psychological difficulty is associated another, consisting of the fact that the physiological processes with which the sensation of time is connected are still less known, lie deeper, and are more thoroughly concealed than the processes corresponding to the other sensations. Our analysis, therefore, must confine itself chiefly to the psychological side, without approaching the question from its physical aspect, as is possible, in part at least, in the provinces of the other senses.

It is scarcely necessary to lay special emphasis on the important part played in our psychical life by the temporal ordering of the elements. The temporal order is even more important than the spatial. Reversal of the temporal order is even more destructive of a process than is the reversal of an object in space by turning it upside down; reverse the temporal order, and an experience becomes something other than itself, something quite new. This is why the words of a speech or a poem are reproduced only in the order in which they were experienced and not in the reverse order as well, in which they would generally have a quite different meaning, or no meaning at all. If the whole acoustic sequence is reversed by saying something backwards, or by making a phonograph work backwards, we do not even recognize any longer the words that are the component parts of the speech. Definite memories are connected only with the definite sequence of sounds in which a word occurs, and it is only when the memories are evoked in a definite order corresponding to the word-sequence, that they combine together to produce a definite meaning. [2] But a sequence of notes too, a simple melody in which habit and association in any case play a very small part, becomes unrecognizable if it is temporally reversed. As regards even very elementary representations and sensations, their temporal sequence forms part of the memory image of them.

If we conceive time as a sensation, it seems less strange that, in a series passing in the order $A\ B\ C\ D\ E$, any member, C for instance, should call up to the memory only the members that follow it, and not those that precede. The memory-image of a building does not arise with the roof turned downwards. But, for the rest, it does not seem to be a matter of indifference whether the organ B is excited after an organ A, or *vice versa*. There is probably a physiological problem concealed here, which would require to be solved before we can fully understand the fundamental psychological fact of the lapse of reproduced series in one determined direction. [3] It is possible that this fact is connected with the fact that an excitation propagates itself along entirely different paths according to the point at which it first enters into the organism, in the way in which this was explained for physical cases by the considerations on p. 92 and by Fig. 1b. Even when the medium is perfectly homogeneous, if two excitations in it, starting from two distant points, spread uniformly, they will more nearly coincide at that one of the two points which was excited later. Thus, even in the simplest cases, the order of stimulation cannot be a matter of indifference.

Let a note D follow a note C. The impression is quite different from what it would be if C followed D. The cause of this is chiefly the notes themselves, and their reciprocal action. For if the pause between the two notes is made sufficiently long, it is possible that we shall no longer distinguish the two cases. Something analogous can be observed with sequences of colors, and in general with sensations of any kind. But if a note A is followed by a color or a smell B, we always know that B has followed A, and our estimate of the pause between A and B is practically not influenced at all by the quality of A and B. There must therefore be a further process, which is unaffected by variation in the quality of sensation, which is quite independent of the quality of sensation, and by means of which we estimate time. It is possible, indeed, to make a sort of rhythm out of entirely heterogeneous sensations, such as sounds, colors and impressions of touch.

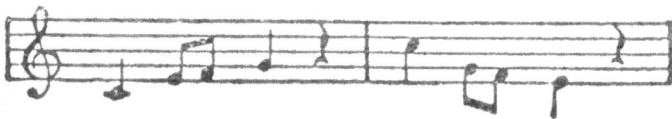

2.

That a definite, specific time-sensation exists, appears to me beyond all doubt. The rhythmical identity of the two adjoined measures, in which the sequence of the notes is quite different, is immediately recognized. We have not to do here with a matter of the understanding or of reflexion, but with one of sensation. In the same manner that bodies of different colors may possess the same spatial form, so here we have two tonal entities which, acoustically, are differently colored, but possess the same temporal form. As in the one case we pick out by an immediate act of feeling the identical components of the space-sensation, so here we immediately detect the identical components of the time-sensation, or the sameness of the rhythm.

It is of course only for small times that I hold that there is an immediate sensation of time. We judge and estimate longer times by remembering the processes that took place in them, that is to say, by splitting them up into the smaller parts of which we had an immediate sensation.

3.

On hearing a number of strokes of a bell, which are exactly alike acoustically, I discriminate between the first, second, third, and so on. Is it perhaps the accompanying thoughts, or other accidental sensations, with which the strokes of the bell happen to be associated, that produce these distinguishing marks? I do not believe that any one will seriously uphold this view. How uncertain and unreliable, if this view were true, would our measurement of time be! What would become of it if that accidental background of thought and sensation should suddenly vanish from memory?

While I am reflecting upon something, the clock strikes, but I give no heed to it. After it has finished striking, it may be of importance to me to count the strokes. And as a fact, there arise in my memory distinctly one, two, three, four strokes. 1 give here my whole attention to this recollection, and by this means the subject on which I was reflecting during the striking of the clock, for the moment completely vanishes from me. The supposed background against which I could note the strokes of the bell, is now wanting to me. By what mark, then, do I distinguish the *second* stroke from the *first?* Why do I not regard all the strokes, which in other respects are identical, as one? Because each is connected for me with a special time-sensation which starts up into consciousness along with it. In like manner, I distinguish a memory-image from a creation of fancy by a specific time-sensation which is not that of the present moment.

4.

Since, so long as we are conscious, time-sensation is always present, it is probable that it is connected with the organic consumption necessarily associated with consciousness, - that we feel the work of attention as time. During any severe effort of attention time is long to us, during easy employment short. When we are in a dull. state, hardly noticing our surroundings, the hours pass rapidly away. When our attention is completely exhausted, we sleep. In dreamless sleep, the sensation of time is lacking. When profound sleep intervenes, yesterday is connected with to-day only by an intellectual bond, apart from the feeling common to both that remains the same.

I have already on a former occasion referred to the apparent difference of the ways in which animals of different sizes measure time. [4] But the measurement of time seems to change with age as well. How short the days seem to me now in comparison with the days of my youth! And in my youth I used to watch an astronomical clock that struck the seconds; when I think of that clock now, the second-stroke seems to be appreciably accelerated. I cannot shake off the impression that my physiological time-unit has become larger.

The fatiguing of the organ of consciousness goes on continually in waking hours, and the labor of attention increases just as continuously. The sensations connected with greater expenditure of attention appear to us to happen later.

Normal as well as abnormal psychical events appear to accord with this conception. Since the attention cannot be fixed upon two different sense-organs at once, the sensations of two organs can never occur together and yet be accompanied by an absolutely identical effort of attention. Hence, the one appears later than the other. Something analogous to the so-called personal equation of astronomers, having its ground in analogous facts, is also frequently observed in the same sense-province. It is a well-known fact that an optical impression which arises physically later may yet, under certain circumstances, appear to occur earlier. It sometimes happens, for example, that a surgeon, in bleeding, first sees the blood spirt out and afterwards his lancet enter. [5] Dvorak has shown, [6] in a series of experiments which he carried out at my desire, years ago, that this relation may be produced at will, the object on which the attention is centred appearing (even when it is really from 1/8 to 1/6 of a second later) earlier than that indirectly seen. It is quite possible that the familiar experience of the surgeon may find its explanation in this fact. The time which the attention requires to turn from one place at which it is occupied, to another, is shown in the following experiment instituted by me. [7] Two bright red squares measuring two centimeters across and situated on a black background eight centimeters apart, are illuminated in a perfectly dark room by an electric spark concealed from the eye. The square directly seen appears red, but that indirectly seen appears green, - and often quite intensely so. The retarded attention finds the indirectly seen square when it is already in the stage of Purkinje's positive after-image. A Geissler's tube with two bright red spots at a short

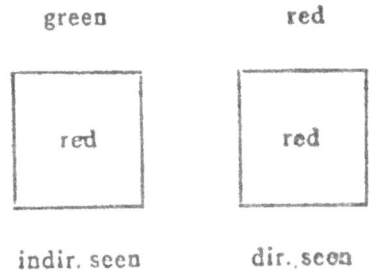

Fig. 34.

146

distance from one another, exhibits, on the passage of a single discharge, the same phenomenon. [8]

The reader must be referred for details to Dvorak's paper. Of particular interest are his experiments on the stereoscopic (binocular) combination of non-simultaneous impressions. [9] More recently Sandford [10] and Münsterberg [11] have carried out experiments of this kind.

5.

I have repeatedly observed an interesting phenomenon which should be cited here. I have been sitting in my room, absorbed in work, while in an adjacent room experiments in explosions were being carried on. It regularly occurred that I shrank back startled, *before* I heard the report.

Since the attention is especially inert in dreams, naturally the most peculiar anachronisms occur in this state, as every one has doubtless observed. For instance, we dream of a man who rushes at us and shoots, awake suddenly, and perceive the object which, by its fall, has produced the entire dream. Now there is nothing absurd in assuming that the acoustic stimulus enters simultaneously different nerve-tracks and is met there by the attention in some inverted order, just as, in the case above mentioned, I perceived first the general excitation and afterwards the report of the explosion. But in many cases it is undoubtedly a sufficient explanation to assume the interweaving of a sensation with the framework of a dream already present.

6.

If organic consumption, or, for that matter, the accumulation of fatigue-material were *immediately* felt, we might logically expect a reversal of time in dreams. The difficulty disappears if consumption and restitution are regarded as heterodromous processes in Pauli's sense (see Ch. IV.). The eccentricities of dreams may all be accounted for by the fact that many sensations and representations do not enter consciousness at all, while others enter with too much difficulty and too late. Inertia of association is a fundamental feature of dreams. The intellect often sleeps only in part. We converse very sensibly, in dreams, with persons long dead, but with no recollection of their death. I speak to a friend of a third person, and this friend is himself the third person of whom I was speaking. We reflect, in the dream-state, concerning dreams, and recognize them as such by their eccentricities, which then at once cease to disturb us. I once dreamed very vividly of a mill. The water flowed downwards, in a sloping channel, away from the mill, and close by, in just such another channel, upwards to the mill. I was not at all disturbed by the contradiction. - At a time when much engrossed with the subject of space-sensation, I dreamed of a walk in the woods. Suddenly I noticed the defective perspective displacement of the trees, and by this recognized that I was dreaming. The missing displacements, however, were immediately supplied. - Again, while dreaming, I saw in my laboratory a beaker filled with water, in which a candle was serenely burning. "Where does it get its oxygen from?" I thought. "It is absorbed in the water," was the answer. "Where do the gases produced in the combustion go to?" The bubbles from the flame mounted upwards in the water, and I was satisfied. W. Robert [12] has made the excellent observation that it is principally perceptions and thoughts, which owing to some interruption we have been unable to carry to a conclusion during the day, of which the thread is taken up in dreams. And as a matter of fact we frequently draw the elements of our dreams from the events of the preceding day. Thus I used to be able to refer, with almost complete certainty, the dream about the light in the water to a certain experiment in my lectures with an electric carbon-light under water, [13] and the dream about the mill to my experiments with the apparatus described (Fig. 19) above. Visual hallucinations play the principal part in my dreams. I have acoustic dreams less often, though I clearly hear conversations, the sound of bells, and music in my dreams. [14] Every sense, even the sense of taste, can come into play in dreams, though some more rarely than others. Since reflex excitability is greatly heightened in the dream-state, and the conscience on the other hand very much weakened owing to the inertia of association, one is capable of almost any crime in dreams, and at the stage of waking may go through the acutest torments. Anyone who allows such experiences to affect him, must entertain grave doubts as to the rightness of our method of exercising justice, which consists in making good one misery by a second, the second being added to the first by means of a process that is revolting because deliberate, cruel, and solemn.

I should not like to let this opportunity slip of recommending to the reader the excellent book of M. de Manacéine. [15] What was said above as to the inadequacy of temporary associations as an explanation of psychical life (see pp. 201, 239, 240) holds for the dream-state also. We have to add that the faintest traces of something which has long been forgotten for the waking consciousness, the slightest disturbances of health and disposition which have to fall into the background during the bustle of the day, can make themselves felt in dreams. In his *Philosophie der Mystik* (1885, p. 123) Du Prel poetically compares this process with the way in which the faintly glimmering starry firmament becomes visible when the sun has set. This book contains many passages of remarkable and profound insight. The man of science, in particular, whose critical sense is directed towards the nearest practicable object of research, reads it with pleasure and profit, without allowing himself to be led astray by the author's inclination towards the fantastic, the miraculous and the extraordinary.

7.

If time-sensation is connected with the growth of organic consumption or with the equally continuous growth of the effort following upon attention, then we can understand why physiological time is not reversible, any more than physical

time, but moves only in one direction. As long as we are in the waking state, consumption and the labor of attention can only increase, not diminish. The two accompanying bars of music, which present a symmetry to the eye and to the understanding, show nothing of the sort as regards the sensation of time. In the province of rhythm, and of time in general, there is no symmetry.

8.

It is perhaps an obvious and natural, though still an imperfect conception, to regard the "organ of consciousness" as capable, in a small degree, of all the specific energies, of which each sense-organ is able only to display a few. Hence the shadowy and evanescent character of representation as compared with sensation, through which it must be constantly nourished and re-freshed. Hence also the capacity of the organ of consciousness to serve as a bridge of connexion between all sensations and memories. With every specific energy of the organ of consciousness, we should then have to conceive still another particular energy, the sensation of time, associated, so that none of the former could be excited without the latter. Should this new energy appear physiologically superfluous and only invented ad hoc, we might at once assign to it an important physiological function. What if this energy kept up the flow of blood that nourishes the brain-parts in their work, guided this current to its destination, and regulated it? Our conception of attention and of time-sensation would then receive a very material basis. The fact that there is only one cohering time, too, would become intelligible, since the partial attention given to one sense is always drawn from the total attention, and is determined by it.

Such a theory is strongly suggested by Mosso's work on plethysmography and by his observations on the circulation of the blood in the brain. [16] William James gives a cautious assent to this conjecture. [17] James indicates that it would be desirable to put it into a more definite and detailed form, but I have unfortunately not been able to do this.

9.

In listening to a number of similar strokes from a bell, we can distinguish each from the others in memory and also count them in memory, provided they are few in number. If the number is large, however, we distinguish the last ones from one another, but not the first. In this case, if we would not make a mistake, we must count them immediately upon their being sounded, that is, we must voluntarily connect each stroke with an ordinal symbol. The phenomenon is perfectly analogous to that which we observe in the province of the space-sense, and is to be explained on the same principle. In walking forwards, we have a distinct sensation that we are moving away from a starting-point, but the physiological measure of this removal is not proportional to the geometrical. In the same manner, elapsed physiological time is subject to perspectival contraction, its single elements becoming less and less distinguishable. [18]

10.

If a special time-sensation exists, it goes without saying that the identity of two rhythms will be immediately recognized. But we must not leave the fact unnoticed that two rhythms which are the same physically may appear very different physiologically, just as the same space-figure by change of position may give rise to different physiological space-forms. The rhythm represented by the following notes, for example, appears quite different according as we regard the short thick, or the long thin vertical lines, or the dotted lines, as marking the bars. Evidently this is connected with the fact that the attention (guided by the accent) sets in at 1, 2, or 3, that is, that the sensations of time corresponding to the successive beats are compared with different initial sensations.

When all the times of a rhythm are prolonged or shortened, a similar rhythm arises, which, however, can only be felt as similar when the prolongation or shortening does not exceed the limit imposed by the immediate sensation of time.

The rhythm represented in the following diagram appears physiologically similar to the preceding, but only when similarly-marked bars are taken in the two - that is, when the attention sets in at homologous points of time. Two physical time-constructions may be termed similar when all the parts of the one stand in the same relation to one another as do the homologous parts of the other. But physiological similarity makes its appearance only when the above condition is likewise fulfilled. Furthermore, so far as I am able to judge, we recognize the identity of the time-ratios of two rhythms only when the same are capable of being represented by very small whole numbers. Thus we really notice immediately, only the identity or non-identity of two times, and, in the latter case, we recognize the ratio of the two only by the fact that one part is exactly contained in the other. Herewith we have an explanation of the fact that, in marking time, the time is always divided into absolutely equal parts. [19]

The conjecture thus forces itself upon us, that the sensation of time is closely connected with periodically or rhythmically repeated processes. But it is scarcely capable of being proved, though the attempt has been made in some quarters, that the measurement of time in general is based on breathing or on the pulse. These questions, however, are by no means simple. Many processes, of course, take place rhythmically in the bodies of animals, without it being possible for us to attribute to them any particular sense for time, rhythm or beat. When a pair of horses is driven past my house, I can hear for a long time the coincidence and alternation of the hoof-beats of the two horses fading away into the distance in regular periods. Thus each horse keeps to its own time without troubling about that of the other horse, and without adapting itself to the other. Two men harnessed together would find this almost intolerable. Wallaschek mentions the deficiency of the sense of time in horses, and also the difficulty of keeping up the appearance of it in circus performances. It can scarcely be upon the coarser bodily processes that the feeling for time is immediately based. Probably it must be referred rather to a superior psychical sensibility, in virtue of which a trifling psychical circumstance determines the attention to notice an otherwise indifferent process. But when processes that keep time are carefully observed, and such observation always involves a certain amount of cooperation and imitation, the psychical functions, and finally the coarser bodily functions themselves also, then become adapted to the time. [20]

Dr. R. Wlassak has communicated to me in conversation a remark which I will reproduce in his own words:

"When the sensations are connected with a vivid emotional coloring, time-values are always markedly diminished; this fact accords with the hypothesis that the sensation of time depends on organic consumption. The rule holds both for stretches of time that are filled with strongly pleasurable sensations and for those filled with unpleasant sensations. On the other hand, the sensations that oscillate round the indifference-values of emotional coloring are connected with relatively indistinct sensations of time. These facts indicate that the nervous processes belonging to the sensations of time and to the emotions respectively offer certain analogies.

In point of fact, all attempts to frame a physiological theory of the emotions bring the emotions into relation with consumption; as is done, for instance, in Meynert's or in Avenarius' theory of the emotions."

[1] The position which I here take differs only slightly from that of my "Untersuchungen über den Zeitsinn des Ohres," *Sitzber. d. Wiener Akademie,* Vol. LI., 1865. Into the details of these earlier experiments, begun in 1860, I shall not enter again here. Nor can I here discuss the plentiful material which has resulted from the works of Meumann, Münsterberg, Schumann, Nichols, Hermann, and others. Cp. Scripture, *The New Psychology,* London, 1897, p. 170. For a supplementary discussion, see my *Erkenntnis und Irrtum,* 1905, pp. 415 sqq.

[2] Cp. R. Wallaschek, *Psychologie und Pathologie der Vorstellung,* Leipzig, 1905, especially the chapter on "The Whole and its Parts," pp. 15 sqq.

[3] Perhaps the nervous elements are not merely endowed with a permanent innate faculty of polar orientation, such as is made probable by the backward direction of the wave in the intestines and the musculature of snakes, and by galvanotropic phenomena, but perhaps they are also capable of a temporarily acquired polarity, as manifested in the inclusion of the time-series in memory, in practice, etc. Cp. Loeb and Maxwell, *Zur Theorie des Galvanotropismus,* Pflüger's *Archiv,* Vol. LXIII. p. 121; Loeb, *Vergleichende Gehirnphysiologie,* pp. 108 sqq.

[4] *Zeitsinn des Ohres,* p. 17.

[5] Compare Fechner, *Psychophysik,* Leipzig, 1860, Vol. II., p. 433.

[6] Dvorak, "Ueber Analoga der persönlichen Differenz zwischen beiden Augen und den Netzhautstellen desselben Auges," *Sitzber. d. königl, böhm. Gesellschaft der Wissenschaften* (*Math-naturw. Classe*), March 8, 1872.

[7] Communicated by Dvorak, *loc. cit.*

[8] G. Heymans could not succeed at first in this latter experiment, but has subsequently convinced himself of the correctness of my statement.

[9] *Op. cit.,* p. 2.

[10] Sandford, *American Journal of Psychology,* 1894, Vol. VI., p. 576.

[11] Münsterberg, *Psychological Review,* 1894, Vol. I., p. 56.

[12] W. Robert, *Ueber den Traum,* Hamburg, 1886.

[13] *Prinzipien der Warmelehre,* 2nd ed., 1900, p. 444.

[14] Wallaschek, "Das musikalische Gedächtnis," *Vierteljahrsschrift für Musikwissenschaft*, 1882, p. 204.

[15] M. de Manacéine, *Sleep, its Physiology*, etc., London, 1897.

[16] Mosso, *Kreislauf des Blutes im Gehirn,* Leipzig, 1881. Cf. also Kornfeld, *Ueber die Beziehung von Atmung und Kreislauf zur geistigen Arbeit,* Brünn, 1896.

[17] W. James, *Psychology,* Vol. I., p. 635.

[18] Cp. Ch. 7.

[19] The similarity of space-figures would be felt, according to this theory, much more immediately than the similarity of rhythms.

[20] Wallaschek, *Anfänge der Tonkunst,* Leipzig, 1903, pp. 270, 271. This book, a profusely illustrated German edition of an English book by the same author (*Primitive Music,* London, 1903), contains many very valuable observations on the questions discussed in this and the following chapters.

XIII. Sensations of Tone

[1]

1.

As regards tone-sensations, also, we are restricted mainly to psychological analysis. As before, the beginning of an investigation is all we can offer.

Among the sensations of tone possessing greatest importance for us are those excited by the human voice, as utterances of pleasure and pain, of expressions of the will, and of the communication of thoughts by speech, etc. Nor can there be any doubt that the voice and the organ of hearing bear a close relation to each other. The simplest and distinctest form in which sensations of tone reveal their remarkable characteristics is music. Will, emotion, the expression of sound, and the sensation of sound, have certainly a strong physiological connexion. There is a good deal of truth in the remark of Schopenhauer [2] that music represents the will, and in fact generally in the designation of music as a language of emotion; although this is scarcely the whole truth.

2.

Following the precedent of Darwin, H. Berg has, to put it shortly, attempted to derive music from the amatory cries of monkeys. [3] We should be blind not to recognize the service rendered and enlightenment conveyed by the work of Darwin and Berg. Even at the present day, music has power to touch sexual chords, and is, as a fact, widely made use of in courtship. But as to the question wherein consists the agreeable quality of music, Berg makes no satisfactory answer. And seeing that in musical theory he adopts Helmholtz's position of the avoidance of beats and assumes that the males who howled least disagreeably received the preference, we may be justified in wondering why the most intelligent of these animals were not prompted to maintain silence altogether.

The importance of tracing the connexion of a given biological phenomenon with the preservation of the species, and of indicating its phylogenetic origin, cannot be underrated. But we must not imagine that in having accomplished this we have solved all the problems connected with the phenomenon. Surely no one will think of explaining the element of pleasure in the specific sexual sensation by showing its connexion with the preservation of the species. We should be more likely to acknowledge that the species is preserved because the sexual sensation is pleasurable. Although music may actually remind our organism of the courtship of distant progenitors, it must, if it was ever used for wooing, have contained at the start some positive agreeable quality, which, to be sure, may be reinforced at the present time by that memory. To take an analogous case from individual life, the smell of an oil-lamp as it goes out almost always agreeably reminds me of the magic lantern which excited my wonder as a child. Yet in itself the smell of the lamp is none the less disgusting for this reason. Nor does the man who is reminded, by the scent of roses, of a pleasant experience, believe, on this account, that the scent was not previously agreeable. It has only gained by the association. [4] And if the view in question cannot sufficiently explain the agreeable quality of music *per se,* it assuredly can contribute still less to the solution of special questions, as, for instance, why, in a given case, a fourth is preferred to a fifth.

3.

A rather one sided view of the sensations of tone would be obtained if we were to consider only the province of speech and music. Sensations of tone are not only a means of communicating ideas, of expressing pleasure and pain, of discriminating between the voices of men, women, and children; they are not mere signs of the exertion or passion experienced by the person speaking or calling; they also constitute the means by which we distinguish between large and small bodies when sounding, between the tread of large and small animals. The highest tones, the very ones which the vocal organs of man cannot produce, presumably are of extreme importance for the determination of the direction from which sounds proceed. [5] In fact, it is more than likely that these latter functions of sensations of tone antedated, in the animal world, by a long period, those which merely perform a part in the social life of animals. By inclining a piece of cardboard in front of the ear, anyone can convince himself that it is only those noises which contain very high tones, such as the rustling and hissing of a gas flame, of a steam kettle, or of a waterfall, that are modified by reflexion according to the position of the cardboard, and that deep tones remain entirely uninfluenced. This shows that it is only in virtue of their effect on high tones that the two ear-conchs can be used as indicators of direction. [6]

4.

There is no one but will cheerfully acknowledge the decided advance effected by Helmholtz in the analysis of auditory sensations, [7] following on the important works of his predecessors, Sauveur, Rameau, R. Smith, Young, Ohm, and others. [8] Following his principles, we recognize in noises combinations of musical tones, of which the number, pitch, and intensity vary with the time. In compound musical sounds, or clangs, we generally hear, along with the fundamental n, the over-tones or partial tones" $2n$, $3n$, $4n$, etc., each of which corresponds to simple pendular vibrations. If two such musical sounds, to the fundamentals of which the rates of vibration n and m correspond, be melodically or harmonically combined, there may result, if certain relations of n and m are satisfied, [9] a partial coincidence of the harmonics, whereby in the first case the relationship of the two sounds is rendered perceptible, and in the second a diminution of beats is effected. All this cannot be disputed, although it may not be considered exhaustive.

We may also give our assent to Helmholtz's physiological theory of audition. The facts observed on the simultaneous sounding of simple notes make it highly probable that there exists, corresponding to the series of vibration-rates, a series of terminal nervous organs, so that for all the different rates of vibration there are different end-organs, each of which responds to only a few, closely adjacent rates of vibration. On the other hand, Helmholtz's physical theories as to the function of the labyrinth have proved untenable. I shall return to this point.

5.

If we assume with Helmholtz that all noises admit of being resolved into sensations of tone varying in duration, it seems superfluous to seek further for a special auditive organ for noises, and Helmholtz himself soon gave up so inconsequent a procedure. A long time ago (in the winter of 1872-73) I took up the question of the relation of noises (especially that of sharp reports) to musical tones, and found that every degree of transition between the two may be demonstrated. A tone of one hundred and twenty-eight full vibrations, heard through a small radial slit in a slowly revolving disc, contracts, when its duration is reduced to from two to three vibrations, to a short, sharp concussion (or weak report) of very indistinct pitch, while with from four to five vibrations, the pitch is still perfectly distinct. On the other hand, with sufficient attention, a pitch, though not a very definite one, may be detected in a report even when the latter is produced by an aperiodic motion of the air (the wave of an electric spark, exploding soap-bubbles filled with $2H + O$). We may easily convince ourselves, furthermore, that in a piano from which the damper has been lifted, large exploding bubbles mainly excite to sympathetic vibration the lower strings, while small ones principally affect the higher strings. This fact, it seems clear to me, demonstrates that the same organ may be the mediator of both tone and noise sensation. We must imagine that weak aperiodic movements of the air having short durations excite all, though preferably the small and more mobile end-organs, whilst more powerful and more lasting movements of the air excite the larger and more inert end-organs as well, which from being less damped perform vibrations of greater amplitude and are thus noticed; and furthermore that even in the case of comparatively weak periodic movements of the air, the stimulus appears, by an accumulation of effects, in some definite member of the series of end-organs. [10] The sensation excited by a report of low or high pitch is qualitatively the same as that produced by striking at once a large number of adjacent piano-keys either high or low in the scale, only more intense and of shorter duration. Moreover, in the single excitation produced by a report, the beats connected with periodic intermittent excitations are eliminated.

6.

The work of Helmholtz excited general admiration on its first appearance; but of late years it has been subjected to various critical attacks, and it almost seems to be as much underestimated now as it was originally overestimated. Physicists, physiologists and psychologists have had nearly forty years in which to test the three several sides of the theory, and it would have been a marvel if they had not found out its weak spots. Without making any pretence to completeness, we will now consider the principal critical objections to it, taking first together the objections which have been urged from the physical and physiological side, and, secondly, those of the psychologists.

Helmholtz assumed, for psychological and physical reasons, that the inner ear consists of a system of resonators, which singles out the members of Fourier's series, corresponding to the form of vibration presented, and hears them as partial tones. On this view, the relation between the phases of the partial vibrations can exert no influence on sensation. As against this view, König, [11] an eminent specialist in acoustics, tried to prove that mere displacement of the phases of the partial pendular vibrations causes a change in the sensational impression or "sound-color." L. Hermann, [12] however, succeeded in showing that when the direction of movement of a phonograph is reversed, no change of sound-color results. According to Hermann, the individual sinuous bands of König's wave-siren do not produce any simple tones, and König's conclusions must therefore be based on a mistaken presupposition. [13] This difficulty may therefore be taken as removed.

The phenomena connected with the combination of tones are not so easily explicable on Helmholtz' point of view. Young supposed that beats of sufficient rapidity could themselves be heard as tones that is to say, that they become compound tones. But since it is impossible to excite any resonator by means of beats, to the tempo of which it is tuned, but only by means of tones, it would be impossible, on the resonance theory, to hear any compound tones. Helmholtz therefore postulated at the outset that compound tones must either be explained objectively by means of powerful tones in virtue of the deviation of similar vibrations from linearity, - or subjectively by means of asymmetrical or non-linear conditions of vibration of the resonating parts of the inner ear. Now König [14] failed to prove the existence of objective compound tones, but discovered on the other hand that, even between tones widely removed from one another, there are beats which can invariably be heard as particular tones when the sequence is sufficiently rapid. Hermann [15] detected compound tones with co-operating tones of such feebleness, that the compound tones seem entirely inexplicable on Helmholtz' theory, either objectively or subjectively. Hermann accordingly holds the view, associating himself on this point with König, that the ear re-acts, not only to wave-shaped vibrations, but also, with a sensation determined by the duration of the period, to every kind of periodicity.

The physical resonance-theory seems, at any rate, in its original form, not tenable; but Hermann [16] thinks that it can be replaced by a physiological resonance-theory. I will deal later with this view, as well as with Ewald's new physical theory of audition.

7.

We now turn to the principal objections brought against Helmholtz from psychological points of view. The lack of a positive factor in the explanation of consonance has been very generally felt, the mere absence of beats not being regarded as a sufficient and satisfactory characterization of harmony. Thus A. v. Oettingen [17] feels the want of some expressed positive element characteristic of each interval (p. 30), and refuses to regard the value of an interval as dependent upon the physical accident of the overtones contained in the sounds. He believes that the positive element in question is to be found in the accompanying remembrance of the common fundamental tone (or tonic), as the harmonics of which the composite notes or clangs of the interval have often occurred, or in the accompanying *remembrance* of the common overtone (or phonic) [18] belonging to the two (pp. 40, 47). On the negative side of his criticisms I am in complete agreement with Von Oettingen. But "remembrance" does not quite fill the need of the theory, for consonance and dissonance are not matters of representative activity, but of sensation. My opinion, therefore, is that A. von Oettingen's conception is physiologically inadequate. His enunciation of the principle of duality, however (or of the principle of the tonic and phonic relationship of composite notes), as also his conception of dissonances as indeterminate composite musical sounds admitting of more than one interpretation, appear to me to be valuable and positive services to science. [19]

<h2 align="center">8.</h2>

Stumpf has in various writings criticized the doctrine of Helmholtz with great penetration. [20] He questions, in the first place, the two different definitions which Helmholtz gives of consonance, - the definition by disappearance of beats, and the definition by coincidence of partial tones, - pointing out that the former is inapplicable to and not characteristic of melodic sequence, and the latter inapplicable to and not characteristic of harmonic combination. A pure triple compound note, intermitting according to the nature of the beats, is not a dissonance. On the other hand, examples can be given of the simultaneous sounding of tones far removed from one another, which produce a violent dissonance, although the beats become imperceptible. If two notes of the tuning-fork are distributed one to each ear, the beats do indeed sink very much into the background, but without the distinction between consonance and dissonance becoming any less. Subjectively heard tones, too, such as ringing in the ears, can be experienced as dissonances, of course without the beats being heard. Tones, finally, that are merely represented, also appear as consonant and dissonant, without the representation of beats playing any essential part in the process. The coincidence of the partial tones ultimately disappears, when no overtones are present, without necessarily causing the disappearance of the distinction between dissonance and consonance. I will pass over Stumpf s polemic against the explanation of consonance by means of unconscious counting, - a view which will probably find few supporters. [21] Equally readily will it be admitted that agreeableness is not a sufficiently characteristic property of consonance, since it is a property which under certain circumstances can just as well belong to dissonance.

Stumpf himself finds the characteristic mark of consonance in the fact of two tones when sounded together approximating, sometimes more and sometimes less, to the impression of a single tone. He defines consonance by means of "fusion," harking back, as it were, to views prevalent in antiquity, of which he gives an exhaustive history. [22] Helmholtz also is not unfamiliar with this theory; he discusses it, but thinks that he has given the first correct explanation of the fusion of notes.

Stumpfs statistical experiments show that a fusion of tones takes place in consonance. If two tones are sounded simultaneously, unmusical persons mistake them for a single tone with a frequency in proportion to the extent to which they are consonant. Stumpf does not attempt to conceal the necessity for some further explanation of fusion. If it is similarity that causes tones to fuse, then this must be a different kind of similarity from that on which the sequence of tones in a series depends, since this latter similarity decreases continuously with the distance of the tones from one another. But since such a second relation of similarity appears to him purely hypothetical, he prefers to imagine a physiological explanation of a different kind. He supposes that, when two tones, of which the rates of vibration stand to one another in a comparatively simple ratio, are heard simultaneously, the cerebral processes which take place are connected by a closer relation of specific synergy than when the ratio of the rates of vibration is more complicated. [23] Tones that succeed one another can fuse. Although polyphonous music is preceded historically by homophonous, yet Stumpf considers it probable that even in the case of homophonous music the selection of the scale was guided by experience of the simultaneous hearing of tones. In all essential points it is impossible not to agree with Stumpf's criticism.

9.

I myself, as early as 1863, [24] and also later, [25] had made some critical remarks on the theory of Helmholtz, and in 1866, in a small work [26] which appeared shortly before that of Von Oettingen, I very definitely pointed out some demands which a more perfect theory of the subject would have to satisfy. I developed these remarks in more detail in the first edition of this book (1886).

Let us start from the idea that a series of definitely graduated end-organs exists, the members of which, as the rate of vibration increases, successively yield their maximum response, and let us ascribe to each end-organ its particular (specific) energy. Then there are as many specific energies as there are end -organs, and as many rates of vibration that we distinguish by the sense of hearing.

Further, we not only distinguish between tones, but we also order them in a series. Of three tones of different pitch, we recognize the middle one immediately as such. We feel immediately which rates of vibration lie nearer together and which are further apart. This is readily enough explained for adjacent tones. For, if we represent the vibration-amplitudes of a certain tone symbolically by the ordinates of the curve a b c. Fig. 35, and imagine this curve gradually moved in the direction of the arrow, then, since necessarily several organs always yield simultaneous responses, neighboring tones will always have faint, common excitations. But more distant tones also possess a certain similarity; and even between the highest and lowest tones we can detect a resemblance. Consequently, in accordance with the principle of

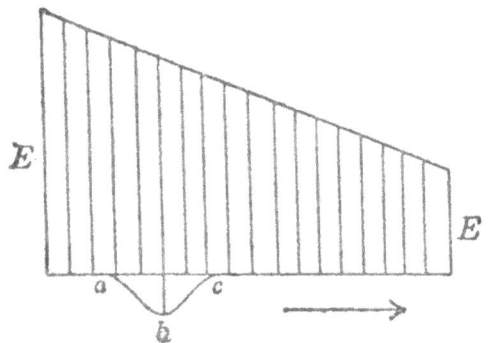

Fig. 35.

158

investigation by which we are guided, we are obliged to assume in all tone-sensations common component parts.

Consequently, there cannot be as many specific energies as there are distinguishable tones. For the understanding of the facts with which we are here concerned, it suffices to assume only two energies, which are excited in different proportions by different rates of vibration. Further complexity of the sensations of tone is not excluded by these facts, but on the contrary is rendered probable by phenomena to be discussed later.

Careful psychological analysis of the tonal series leads immediately to this view. But even supposing we assume a special energy for every end-organ, and reflect that these energies are similar to one another, that is, must contain common component parts, virtually we arrive at the same standpoint. Let us therefore assume, merely in order to have a definite picture before us, that, in the transition from the lowest to the highest rates of vibration, the tonal sensation varies similarly to the color-sensation in passing from pure red to pure yellow, say by the gradual admixture of yellow. We can fully retain, on this view, the idea that there is for every distinguishable rate of vibration a special appropriate end-organ; but in that case not absolutely different energies, but always the same two energies, only in different proportions, are disengaged by the different organs. [27]

10.

How does it happen, now, that so many tones simultaneously sounded are distinguished, and are not fused into a single sensation; or that two tones of different pitch do not blend to a mixed tone of intermediate pitch? The fact that this does not happen, lends a still more definite shape to the conception which we have to form. The case is probably similar to that of a graduated series of mixed reds and yellows situated at different points of space, which are likewise distinguished and do not blend. And in fact, the sensation which ensues when the attention passes from one tone to another is similar to that which accompanies the wandering of the fixation-point in the field of vision. The tonal series occurs in something which is an analogue of space, but is a space of one dimension limited in both directions and exhibiting no symmetry like that, for instance, of a straight line running from right to left in a direction perpendicular to the median plane. It more resembles a vertical right line, or one running from the front to the rear in the median plane. But while colors are not confined to certain points in space, but may move about, which is the reason we so easily separate space-sensations from color-sensations, the case is different with tone-sensations. A particular tone-sensation can occur only at a particular point of the said one-dimensional space, on which the attention must in each case be fixed if the tone-sensation in question is to be distinctly perceived. We may now imagine that different tone-sensations have their origin in different parts of the auditive substance, or that, in addition to the two energies whose ratio determines the timbre of high and deep tones, a third exists, which is similar to a sensation of innervation, and which comes into play in the fixation of tones. Or both conditions might occur together. At present it may be regarded as neither possible nor necessary to come to a conclusion in the matter.

That the province of tone-sensation offers an analogy to space, and to a space having no symmetry, is unconsciously expressed in language. We speak of high tones and deep tones, not of right tones and left tones, although our musical instruments suggest the latter designation as a very natural one.

11.

In one of my earliest publications [28] I supported the view that the fixation of tones was connected with a varying tension of the *tensor tympani*. I am now unable to maintain this view in the light of subsequent observations and experiments which I have made. Nevertheless, the space-analogy does not fall to the ground for this reason; only the appropriate physiological element remains to be discovered. The supposition that the processes in the larynx during singing have something to do with the formation of the tonal series was likewise noticed by me in my work of 1863, but I did not find it tenable. Singing is connected with hearing in too extrinsic and accidental a manner. I can hear and imagine tones far beyond the range of my own voice. In listening to an orchestral performance with all the parts, or in having an hallucination of such a performance, it is impossible for me to think that my understanding of this broad and complicated sound-fabric has been effected by my one larynx, which is, moreover, no very practised singer. I consider the sensations which, in listening to singing, are doubtless occasionally noticed in the larynx, a matter of subsidiary importance, like the pictures of the keys touched which, when I was more in practice, sprang up immediately into my imagination on hearing a performance on the piano or organ. When I imagine music, I always distinctly hear the notes. Music can no more come into being merely through the motor sensations accompanying musical performances, than a deaf man can hear the music by watching the movements of players. I cannot therefore, agree with Stricker on this point. (Cp. Strieker, *Du langage et de la musique,* Paris, 1885).

Different is my opinion with regard to Stricker's views on language. (Cp. Strieker, *Die Sprachvorstellungen,* Vienna, 1880.) It is true that in my own case words of which I think reverberate loudly in my ear. Moreover, I have no doubt that thoughts may be directly excited by the ringing of a housebell, by the whistle of a locomotive, etc., and that small children and even dogs understand words which they cannot repeat. Nevertheless, I have been convinced by Strieker that the ordinary and most familiar, though not the only possible way by which speech is comprehended, is really motor, and that we should be badly off if we were without it. I can cite corroborations of this view from my own experience. I frequently see strangers who are endeavoring to follow my remarks slightly moving their lips. If a person tells me his place of residence and I omit to repeat the street and number of the house after him, I am certain to forget the address, but with the exercise of this precautionary measure, I retain it perfectly in memory. A friend told me recently that he would not read the Indian drama *Urvasi,* because he had great difficulty in spelling out the names, and consequently could not retain them in memory. The dream of the deaf-mute, which Strieker relates, is intelligible only from his point of view. In fact, on calm reflexion this seemingly paradoxical relation is by no means so remarkable. The extent to which our thoughts move in accustomed and routine channels is shown by the surprise produced by witticisms. Good jokes would be more frequent if our minds moved less in ruts. To many the obvious collateral meanings of words never suggest themselves. Who, for example, in using the names Smith, Baker, or Taylor thinks of the occupations designated! To adduce an analogous example from a different field, I may state (cp. Ch. VI.) that I immediately recognize writing reflected in a mirror and accompanying its original, as symmetrically congruent with the latter, although I am not able to read it directly, because of my having learned writing by motor methods, with my right hand. I can also best illustrate by this example why I do not agree with Stricker in regard to music: music is related to speech as ornament is to writing.

12.

I have repeatedly illustrated by experiments, which I will cite again here, the analogy between fixing the eyes on points in space and fixing the attention on tones. One and the same combination of two tones sounds different according as we fix our attention upon the one or the other. Combinations i and 2 in the annexed cut have a perceptibly different character according as we fix our attention on the higher or on the lower note. Persons not able to transfer their attention at will will be helped by having one note sounded later than the other (3, 4). The one sounded last then draws the attention after it. With a little practice it is possible to decompose a chord (as, for instance, 5) into its elements and to hear the constituent tones by themselves (as in 6). These and the following experiments are better and more convincingly carried out upon a physharmonica, on which the notes can be held, than on a piano.

Especially astonishing is the phenomenon produced when .we cause one note of a chord, on which the attention is fixed, to be damped. The attention then passes over to one of the notes nearest to it, which comes out with the distinctness of a note that has just been struck. The impression made by the experiment is quite similar to that which we receive when, absorbed in work, we suddenly hear the regular striking of the clock emerge into distinctness after having entirely vanished from consciousness. In the latter case the entire tonal effect passes the threshold of consciousness, whilst in the former a part is augmented. If in 7, for example, we fix the attention upon the upper note, letting go, successively from above, the keys damping the other notes, the effect obtained is approximately that of 8. If, in 9, we fix the attention on the lowest note, and proceed in the reverse order, we obtain the impression represented in 10. The same chordal sequence sounds quite different according to the part on which the attention is fixed. If, in 11 or 12, I fix my attention on the upper note, the timbre alone appears to be altered. But if, in 11, the attention be fixed upon the bass, the entire acoustic mass will seem to sink in depth; while in 12 it will appear to rise if we regard closely the succession e-f. This makes it quite evident, in fine, that chords act the part of clangs (or compound notes embracing both fundamentals and harmonics). The facts here advanced remind us strongly of the changing impression received when, in observing an ornamental design, the attention is alternately fixed on different points.

We may also recall to mind here the involuntary wandering of the attention which takes place during the continuous and uniform sounding of a note on the harmonium, where if the note lasts several minutes, all the overtones will of themselves successively emerge into full distinctness. [29] The process appears to point to a sort of fatigue for the note on which the attention has long been fixed. This fatigue, moreover, is rendered quite probable by an experiment which I have described at length in another place. [30]

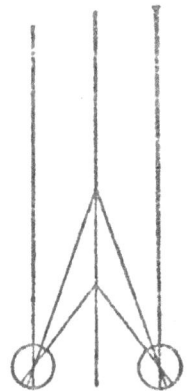

Fig. 36.

The relations in the sphere of tone-sensation, which I we have here been describing, might be illustrated perhaps more palpably by some such parallel as the following. Suppose that our two eyes were capable of only a single movement, and that they could only follow, by changing motions of symmetrical convergence, the points of a horizontal straight line lying in the median plane; and suppose that the nearest point on this line fixed by the eyes were pure red, and the point farthest away, corresponding to the position of parallelism, were pure yellow, while between them lay all inter-mediate shades; then the system of sight-sensations so constructed would quite palpably resemble the relations of the sensations of tone.

13.

On the view hitherto developed, an important fact, which we shall now consider, remains unintelligible, though its explanation is absolutely neces-sary if the theory is to lay any claim to completeness. If two series of tones be begun at two different points on the scale, but be made to maintain through-out the same ratios of vibration, we recognize in both the same melody, by a mere act of sensation, just as readily and immediately as we recognize in two geometrically similar figures, similarly situated, the same form. Like melo-dies, differently situated on the scale, may be termed tonal constructs of like tonal form, or they may be termed similar tonal constructs. It is easy to con-vince oneself that this recognition is not connected exclusively with the em-ployment of ordinary musical intervals or of any comparatively simple ratio of vibration-numbers in common use. If the open strings of a violin, or of any other instrument with more than one string, be "tuned to any disconnected notes we please, and a strip of paper, divided up into any complicated series of ratios, be fastened to the finger-board, we can play the notes indicated in any order (or slide from one to the other), first on one string and then on the others. Now although the resultant sound may have no sense as music, we can recognize the melody as the same on each string. The experiment would not be any more convincing, if we deliberately divided the finger-board into irrational intervals. Indeed in practice the result would be only approximate. The musician could still maintain that he heard intervals that were approxi-mate or intermediate to the familiar musical intervals. Untrained song-birds use the musical intervals only in exceptional cases.

Even in a series of only two tones, the sameness of the ratios of vibration is at once recognized. Thus in the series c-f, d-g, e-a, etc., the notes which have all the same ratios of vibration (3: 4), are immediately recognized as like in-tervals, as fourths. Such is the fact, in its simplest form. The ability to pick out and recognize intervals is the first thing required of the student of music who is desirous of becoming thoroughly familiar with his subject.

In a little work, [31] well worth reading, by E. Kulke, mention is made, bearing on this point, of an original method of instruction by P. Cornelius - a notice which I will now supplement by the following communication made to me orally by Kulke himself. According to Cornelius, it is a great help in the recognition of intervals to make note of particular pieces of music, folk-songs, etc., which begin with these intervals. The overture to *Tannhäuser,* for example, begins with a fourth. If I hear a fourth I at once remark that the tone-sequence might be the beginning of the overture to *Tannhäuser,* and by this means I recognize the interval. In like manner, the overture to *Fidelio* (No. i) may be used as the representative of the third; and so on. This excellent device, which I have put to the test in my lectures on acoustics and have found very effective, apparently complicates matters. One would naturally suppose that it would be easier to make note of an interval than of a melody. Nevertheless, a melody offers a greater hold to memory than does an interval, just as an individual countenance is more easily remarked and associated with a name than is a certain facial angle or a nose. Every one makes note of faces and associates with them names; but Leonardo da Vinci arranged noses in a system.

14.

Just as every interval in a sequence of tones is made perceptible in a characteristic manner, so it is with the harmonic combinations of tones. Every third, every fourth, every major or minor triad has its characteristic color, by which it is recognized independently of the pitch of the fundamental, and independently of the number of beats, which rapidly increases with increasing pitch.

A tuning-fork held before one ear is very feebly heard by the other ear. If two slightly discordant, beating tuning forks are held in front of the same ear, the beats are very distinct. But if one of the forks be placed before one ear, and the other before the other, the beats will be greatly weakened. Two forks of harmonic interval always sound slightly rougher before one ear. But the character of the harmony is preserved when one is placed before each ear. [32] The discord also remains quite perceptible in this experiment. Harmony and discord are, however, not determined by beats alone.

15.

In melodic as well as in harmonic combinations, notes whose rates of vibrations bear to one another some simple ratio, are distinguished (1) by their agreeableness, and (2) by a sensation characteristic of this ratio. As for the agreeable quality, there is no denying that this is partly explained by the coincidence of the overtones, and, in the case of harmonic combination, by the consequent effacement of the beats, resulting always where the ratios of the numbers representing the vibrations satisfy certain definite conditions. But the experienced and unprejudiced student of music is not entirely satisfied with this explanation. He is disturbed by the preponderant role accorded to the accident of acoustic color, and notices that tones further stand to each other in a positive relation of contrast, like colors, except that, in the case of colors, no such definite agreeable relations can be specified.

The fact that a sort of contrast really does exist among tones almost forces itself upon our notice. A smooth, unchanging tone is something very unpleasing and colorless, like a single uniform color enveloping our entire surroundings. A lively effect is produced

only on the addition of a second tone, a second color. In like manner, if we cause a tone gradually to mount in pitch, as in experiments with the siren, all contrast is lost. Contrast exists, however, between tones farther apart, and not merely between those immediately following one another, as the accompanying example will show. Passage 2 sounds quite different after 1 from what it does alone, 3 sounds different from 2, and even 5 different from 4 immediately following 3.

C	c	g	c̄	e↓̿	ḡ	b̄-flat	c̿
n	$2n$	$3n$	$4n$	$5n$	$6n$	$7n$	$8n$

E	e	b	ē↓	g-sharp	b̄	d̿	e̿
m	$2m$	$3m$	$4m$	$5m$	$6m$	$7m$	$8m$

F	f	c̄	f̄✝↓	a	c̿	e-flat̿	f̿
n	$2n$	$3n$	$4n$	$5n$	$6n$	$7n$	$8n$

A	a	e	ā✝↓	c-sharp̿	e̿	g̿	a̿
m	$2m$	$3m$	$4m$	$5m$	$6m$	$7m$	$8m$

16.

Let us turn now to the second point, the characteristic sensation corresponding to each interval, and ask if this can be explained on our present theory. If a fundamental n be melodically or harmonically combined with its third m, the fifth harmonic of the first note ($5n$) will coincide with the fourth of the second note ($4m$). This, according to the theory of Helmholtz, is the common feature characterizing all third combinations. If I combine the notes C and E, or F and A, representing their harmonics in the above table, then, as a fact, in the one case the harmonics marked ↓ and in the other those marked ✝ coincide; and in both cases the coincidence is between the fifth harmonic of the lower and the fourth harmonic of the higher note. Be it noted, however, that this common feature exists solely for the *understanding*, being the result of a purely physical and intellectual analysis, and has nothing to do with *sensation*. For sensation the real coincidence in the first case is between the ē's, and in the second between the ā's, which are entirely different notes. On the assumption that there exists for every distinguishable rate of vibration an appropriate specific energy, we are obliged, more than on any other theory, to ask where is the common component of sensation hidden that characterizes every third combination?

165

I must insist on this distinction of mine not being regarded as a piece of pedantic hair-splitting. I propounded the question involving it about twenty years ago, at the same time with my question as to wherein physiological similarity of form, as distinguished from geometrical, consisted; and the former is not a whit more unnecessary than was the latter, of which the superfluity, too, in the issue, was disproved. If we are to allow a physical or mathematical characteristic of the tierce-interval to stand as a mark of the tierce-sensation, then we should content ourselves, as Euler did, [33] with the coincidence of every fourth and fifth vibration a conception which was, after all, not so bad, as long as it could be believed that sound continued its course in the nerve-tracts, also, as periodic motion, a view which even A. Seebeck [34] (*Pogg. Ann.,* Vol. LXVIII.) regarded as possible. With regard to this particular point, Helmholtz's coincidence of $5n$ and $4m$ is in no respect less symbolical and does not offer greater enlightenment.

17.

So far I have presented my arguments with the conviction that I should not find it necessary to make a single retrograde step of importance. This feeling does not accompany me in the same measure in the development of the following hypothesis, which, in all its essential features, was suggested to me a long time ago. Yet the hypothesis may at least serve to clear up and illustrate, from the positive side also, the requirement which I believe a more complete theory of tone-sensations is bound to meet. I will first expound my view in the form in which it appeared in the first edition of this book.

Let us suppose that it is an extremely important vital condition for an animal of simple organization to perceive slight periodic motions of the medium in which it lives. If (owing to the relatively excessive size of its organs, and its consequent lack of receptivity for such rapid changes) change of attention is too sluggish, and the period of the oscillations is too short, or their amplitude too small, to permit the single phases of the excitation to enter consciousness, it may nevertheless be possible under certain conditions for the animal to perceive the accumulated sensation-effects of the oscillatory stimulus. The organ of hearing will outstrip the organ of touch. [35] Now an end-organ capable of vibration (say an auditory cilium) responds, by virtue of its physical qualities, not to every rate of vibration, nor to one only, but ordinarily to several, at a considerable distance apart. [36] Therefore, as soon as the whole continuum of rates of vibration between certain limits becomes of importance for the animal, a small number of end organs no longer suffices, but the need of a whole series of such organs of graduated capacity arises. At first the organ of Corti and the basilar membrane were regarded by Helmholtz as such a system.

It can hardly be expected, however, that a member of such a system will respond to only one rate of vibration. On the contrary, we should expect that it would respond with enfeebled but graduated intensity (perhaps from being divided by nodes) to the rates of vibration $2n$, $3n$, $4n$, etc., as also to the rates of vibration $n/2$, $n/3$, $n/4$, etc. Inasmuch as the assumption of a special energy for each rate of vibration has proved untenable, we may imagine, agreeably to what has been said above, that in the first place, only two sensation-energies, say, Dull (D) and Clear (C), are excited. The resultant sensation we will represent symbolically (somewhat as we do in mixed colors) by $pD + qC$; or, making $p + q = 1$ and regarding q as a function $f(n)$ of the rate of vibration, [37] by $[1 - f(n)] D + f(n) C$. The sensation arising will now correspond to the number of the vibrations of the oscillatory stimulus, to whatever member of the series of end-organs the stimulus may be applied. And consequently our earlier view will not be materially disturbed by the new hypothesis. For, since the member R_n responds most powerfully to n, and only in a much more enfeebled degree to $2n$, $3n$, or to $n/2$, $n/3$, R_n vibrating with n even in case of an aperiodic impulse, therefore the sensation $[1 - f(n)] D + f(n) C$ will still be predominantly associated with R_n.

Well-attested cases of double hearing (cp. Stumpf, *loc. cit.*, p. 266 et seq.) point forcibly to the conclusion that the ratios in which D and C are disengaged are dependent upon the end-organ, and not upon the rate of vibration a conclusion which would also not affect our theory.

A member R_n, accordingly, responds powerfully to n, and also, though more weakly, to $2n$, $3n$,.... $n/2$, $n/3$.... with the sensations belonging to these rates of vibration. It is, however, extremely improbable that exactly the same sensation is excited whether R_n responds to n, or whether R_{n2} responds to n. On the contrary, it is probable that every time the members of the series of organs respond to a partial tone, the sensation receives a weak supplementary coloring, which we will represent symbolically, for the fundamental tone by Z_1, for the overtones by Z_2, Z_3, ..., and for the undertones by $Z_{1/2}$ $Z_{1/3}$,... On this supposition, sensations of tone would be somewhat richer in composition than would follow from the formula $[1 - f(n)] D + f(n) C$. The sensations which the series of end-organs, as excited by the fundamentals, yields, constitute a province with the supplementary coloring Z_1, the excitation of the same series by the first overtone yields a special province of sensation with the supplementary coloring Z_2, etc. The Z's may be either unchanging elements, or may themselves, again, consist of two components, U and V, and form series representable by $[1 - f(n)] U + f(n) V$. But at present the decision on this last point is immaterial.

It is true that the physiological elements Z_1, Z_2,... have yet to be found. Yet the very perception that they have to be sought seems to me of importance. Let us see what form the province of tone-sensations would take on if we regarded Z_1, Z_2,.... as given.

We will take as example a melodic or harmonic major-third combination, whose rates of vibration are $n = 4p$ and $m = \$5p$; the lowest of the overtones common to the two is $5n = 4m = 20p$, the highest of the undertones common to the two is p. Then we obtain the table on the following page.

Thus in the third combination, the supplementary sensations Z4, Z5, and Z¼ Z1/5, which are characteristic of the third, make their appearance even when the notes contain no overtones, while the former (Z4, Z5) are strengthened when, either in the open air or at least in the ear, overtones do occur. The diagram may be easily generalized to include any interval. [38]

	The members of the series of end-organs:	$R\,p$	$R\,4p$	$R\,5p$	$R\,20\,p$
When the notes $4p$ and $5p$ do not contain overtones.	respond to the rates of vibration:	$4p,\ 5p$	$4p$	$5p$	$4p = \dfrac{20\,p}{5}$ $5p = \dfrac{20\,p}{4}$
	with the supplementary sensations:	$Z_4,\ Z_5$	Z_1	Z_1	$Z^{\,3}_{\frac{5}{8}},\ Z_{\frac{1}{4}}$
When the notes $4p$ and $5p$ contain overtones,	they also respond to the rates of vibration:		$20p = 5\,(4p)$	$20p = 4\,(5p)$	
	with the supplementary sensations:		Z_5	Z_4	

These supplementary colorings, though scarcely noticeable in single tones, or in running continuously through the scale, will accordingly become conspicuous in combinations of tones having certain ratios of rate of vibration, just as the contrasts of faintly colored, almost white lights become vivid when such lights are brought together. And, furthermore, the same contrast-colorings always correspond to the same ratios of rate of vibration, no matter what the pitch.

In this manner it is intelligible how tones may receive, by melodic and harmonic combination with others, the most varied colorings, which are wanting to them when singly sounded.

The elements Z_1, Z_2...must not be conceived as unvarying and fixed in number. On the contrary, it is to be supposed that the number of perceptible Z's depends on the organization, on the training of the ear, and on the" attention. According to this conception, the ear does not directly cognize ratios of rate of vibration but only the supplementary colorings conditioned by these ratios. The tonal series symbolically represented by $[1 - f(n)]\,D + f(n)\,C$ is not infinite but limited. Since $f(n)$ may vary between the values 0 and 1, D and C, where they are the sensations corresponding to the lowest and highest tones, are the end-terms of the series. If the number of vibrations sinks considerably below or rises considerably above that of the fundamental of the extreme term of the series, a weak response only will take place, but no alteration of the quality of the sensation. Further, the sensation of the intervals must disappear in the neighborhood of the two limits of hearing; first, because, in general, differences between sensations of tone cease at this point, and, secondly, because at the upper limit the members of the series capable of being excited by the undertones are lacking, while at the lower limit those which react on the overtones are lacking.

Passing in review again the position at which we have arrived, we see that with few exceptions the conclusions reached by Helmholtz may be all retained. Noises and composite sounds may be decomposed into musical tones. To every distinguishable rate of vibration there corresponds a particular nervous end-organ. In place of the numerous specific energies required by this theory, however, we substitute two only, which render the relationship of all tonal sensations intelligible, and which, by the role which we assign to the attention, likewise enable us to keep perceptually distinct, several tones when sounded together. By the hypothesis of the multiple response of the members of the series of end-organs, and by that of supplementary acoustic colorings, the significance of accidental acoustic color is diminished, and we get a glimpse of the direction in which, notably on the ground of musical facts, the positive characteristics of intervals are to be further investigated. Finally, the latter conception supplies Von Oettingen's principle of duality with a foundation which might perhaps commend itself to this investigator himself as preferable to his assumption of "memory"; while at the same time it becomes manifest why the duality cannot be a perfect symmetry.

18.

I have expressly described the theory of the multiple response of the series of end-organs, as well as the theory of supplementary colorings, as a "hypothesis," and I have put forward this hypothesis merely with the object of illustrating the meaning of the postulates resulting from psychological analysis, and of perhaps stimulating others to a more successful attack on the problem. I am therefore not surprised to find that other writers do not agree unreservedly with my attempt. But that this hypothesis is, as Stumpf says, [39] useless and quite unsuited to its purpose, I cannot admit. The coincidence of the supplementary colorings Z_4, Z_5, or $Z_{1/4}$ $Z_{1/5}$, in one and the same nerve, is not merely a physical, but also a psycho-physical fact. It can scarcely be a matter of indifference that the sensation of a mixed coloring is determined by a single element. On the contrary, it seems to me that what I am looking for, namely an explanation of the definite coloring of the intervals, - and also what Stumpf is looking for, - namely an explanation of fusion, - would actually be represented by the partial coincidence which I assume, even without overtones. Stumpf s further assertion that, in the case of notes with overtones, there is no difficulty in understanding, on Helmholtz's theory, the similarity of like intervals, rests upon a misapprehension of my criticism of Helmholtz. It will satisfy no one to be told that overtones of equal strength coincide in the case of two tierces, since what is in question is the qualitative similarity of the sensations. If the recognition of a melodic tierce-interval were immediately intelligible, it would of course be unnecessary to look for any special explanation of why we recognize the harmonic combination of tierces. But inasmuch as Stumpf himself holds that the melodic steps are characterized by the harmonic combination, this view would involve a vicious circle. On my theory also, the fact of melodic and harmonic selection of definite ratios of vibration-numbers leads to the same problem. My hypothesis inclines towards the theory of resonance, and according to Stumpf is to be rejected on that very account. I will presently deal more particularly with this last point.

19.

There has been much discussion of the physical processes involved in audition, and, in particular, of the function of the parts of the middle ear. In spite of this, it would seem that an unprejudiced revision of the physical theory of audition is urgently needed. The question has been raised whether the auditory ossicles vibrate as a whole, or whether the sound-waves pass through them. E. H. Weber decides in favor of the former view, which has been experimentally confirmed by Politzer, while I was probably the first to establish it on a theoretical basis. [40] For if the dimensions of the ossicles in comparison with the length of the sound-waves in question are, as regards their material, very small (as is actually the case), there can be no doubt that practically the same phases of movement occur throughout the whole extent of the ossicles, and that consequently they must move as a whole. It has occurred to some inquirers to transfer the movements of the ossicles to the fluid of the labyrinth. But pathological investigations teach us that, even without the co-operation of the ossicles and the membrana tympani, the hearing may remain quite good, provided that the labyrinth is in order. The ossicles and tympanum seem only to be important when what is in question is the transference to the labyrinth of very faint movements of the air; in that case the reduction of the pressure falling upon the whole surface of the membrana tympani to the small stirrup-footplate seems to be necessary. Otherwise, sound-waves can be carried to the labyrinth by way of the cranial bones also. If sounding bodies, such as tuning-forks, are placed on different parts of the head, it can be shown that the direction of the sound-waves that penetrate into the labyrinth does not play any important part. Again, all dimensions of the sound-perceiving apparatus are so small in comparison with the audible sound-waves, and the velocity of sound in the bones and the labyrinthine fluid is so great, that the whole extent of the labyrinth only contains room for one perceptible wave-phase at any one moment. The result of all this is, that it is not the movements and the direction of movement, but the variations of pressure which arise in the labyrinth almost synchronously, that are to be regarded as the decisive stimulus that excites sensation.

Nevertheless let us consider the movement which can be set up in the labyrinth by the movements of the stirrup-plate. We may first imagine all the soft parts to be taken out, and the whole space bounded by the osseous wall to contain nothing but fluid. The movement that can find room here is a periodical current passing from the oval to the round window and *vice versa*, the form of which, since the velocity of the disturbance vanishes in proportion to the velocity of the sound, will be almost entirely independent of the period. If the surfaces of the two windows are conceived as positive and negative electrodes, and as conducting the fluid, then the lines of the electric current will coincide with the lines of the periodic current. Now this state of things will not suffer any substantial change, if the difference between the specific weight of the soft parts and that of the fluid in which they are immersed is very small. It is the mass of the fluid that is of predominant importance. The fact that particular constructions can', according to the pitch, take on a special local state of vibration in spite of the fluid, need scarcely be discussed. The quantitative relations are here quite different from what they are in the case of strings or membranes in air.

Consequently, Ewald's new theory of audition [41] is, in my opinion, no more tenable than the theory of Helmholtz as to the fibres of Corti and the elective vibrations of the basilar membrane. If in Ewald's experiments a membrane coated with oil shows, when the coat is comparatively thick, no clear divisions any longer, it would completely fail to show any if it were immersed in a fluid, much more if the dimensions were proportionately reduced. We must, however, insist that Ewald's theory is otherwise appropriate in many respects, and that it would offer many advantages. For instance, in the case of harmonic intervals, even when overtones are absent, the membranes display coincidences of the node-lines. Thus this theory seems to fulfil a part of the above mentioned postulates. Unfortunately it is physically inadmissible, quite apart from further difficulties which it is also unable to solve. I need scarcely say that I do not presume to dismiss so admirable and laborious a piece of work in a few words, but at the same time I cannot refrain from stating my objections to it.

Shortly after the publication of the fourth edition of this book, which contained the foregoing passages expressing my doubts as to the vibration of membranes in fluids, Ewald published his experiments with the camera acustica. [42] He immersed in water a delicate membrane, of about the dimensions of the basilar membrane, and succeeded in acoustically setting up in it continuous vibrations with clear nodal divisions corresponding to the pitch. This showed that my conjecture was wrong, and gave me cause to reflect in what point I had been mistaken. I then remembered the very small nodal divisions which, years before, I had myself observed in fluid membranes. [43] I also recalled Friesach's experiments with strings immersed in water, [44] the result of which had been to show that immersion in water acts as an enlargement of the string's mass, since the fluid in the immediate neighborhood of the string accompanies its vibrations, moving synchronously to and fro in extremely short paths. It is therefore quite conceivable that the labyrinthine fluid vibrates to and fro as a whole, and that nevertheless the velocity of propagation in the membrane, which is very much smaller, appears in the labyrinthine fluid in the form of stable vibrations of the membrane. If the existence of such vibrations of the membrane is proved, Ewald's theories gain greatly in value. I should like, further, to refer here to two papers by Stöhr which seem to me to contain the germs of ideas which would repay development. [45]

20.

The difficulty of setting the theory of resonance on a sound physical basis has probably, as it seems to me, been felt by everyone who has studied it, and not least acutely by its originators. But at the same time it was recognized that, if it was given up, the key to the problem of the analysis of sounds and to a clear and simple doctrine of tone-sensations would be lost. Hence the frantic attempts to save the theory of resonance. L. Hermann [46] seems to me to speak very much to the point, when he says that we cannot do without some sort of theory of resonance, but that this need not necessarily be a physical theory, but may also be a physiological theory. We may make, with him, the plausible assumption that the nervous organs *per se* are peculiarly sensitive to stimuli with a definite period. [47] It cannot exactly be elastic forces that impel the organ back to its position of equilibrium, but we may think of a state of equilibrium which is of an electrical or chemical nature, the deviations from it standing to one another in the relation of the positive to the negative sign. And further, there may be a connexion between these organs, by means of which one can act as a stimulus upon another. In this way a reasonable prospect seems to be opened up of making up for the loss of the physical theory. I cannot here attempt to reproduce Hermann's arguments completely and accurately, but must content myself with referring the reader to his paper.

One point, however, we may consider more closely. When two wave-shaped pendular vibrations with vibration-numbers, n, n', co-operate, there then arise beats which may be regarded as a rising and sinking of the tone, n or n' ($n' - n$) times a second. But it is never possible to regard the movement of the air as such a movement as could contain the wave-vibration, that is to say, the tone $n' - n$. Even a physical resonator with the vibration-number $n' - n$ can never be excited by such beats, whether they are fast or slow. Indeed, it is easy to see, when one imagines or draws the course of the beats, that in the long run precisely as many and as strong positive and negative impulses must take place as there are vibrations ($n' - n$) of the resonator. In the first half of the time, also, the impulses are equal to and of the same direction as the impulses in the second half. An effectual summation is therefore excluded. It would only be possible if the resonator could be made more receptive of one kind of impulse than of the other, and more receptive in the first half of the duration of its vibration. We can easily see how this way of considering the matter involves the rejection of Young's explanation of combination tones by means of rapid beats, and leads, on the other hand, to Helmholtz's theory of combination tones, while preserving the theory of resonance. It would seem, however, that the physical relations which Helmholtz had to assume do not exist in the circumstances under which combination tones are heard. But it is quite conceivable that a nervous organ should be unequally receptive of opposite impulses, and likewise should be receptive in different degrees at different stages of its excitation. For an organ does not simply follow the forces that act upon it, but contains a store of energy upon which those forces only act by liberating the energy. In this way Young's mistake, and the presumably unsuccessful attempt of Helmholtz to improve on Young's theory, will have led to an important new point of view.

21.

The theory of Helmholtz as to tone-sensations seemed, when it was first promulgated, to be an admirably complete and classical achievement. Yet, on a fundamental examination, it has not been able to stand against criticism. And this criticism was in no sense captious, as is sufficiently evident from the fact that the attacks of the different critics, in spite of all individual peculiarities, were directed against the same points and took the same direction. The result of all this criticism seems to be that the main problem has been put back almost to the point at which it stood before Helmholtz wrote. The effect of this might be tragic, if it were ever legitimate to look at the matter from the point of view of a single person.

But the achievement of Helmholtz, open to attack though it may be on certain sides, must not be underrated. His work, apart from the positive increase of knowledge which we owe to it, has brought life and movement into the whole question; it has encouraged inquirers to make new experiments, and has provided the stimulus for a mass of new investigations; new prospects have been opened up, and possible ways of going wrong have been definitely closed for ever. New experiments and criticisms are made all the easier by the existence of some positive work from which to take their start.

In thinking that a task, which provides ample work for psychologists and physiologists as well as for physicists, could be mastered, in all its main features, from physical points of view, Helmholtz was doubtless under a delusion. Even those friends and contemporaries of his who, about the middle of the last century, founded the physical school of physiology in co-operation with him, have had to recognize that the fragment of inorganic physics which we have conquered, is far from being the whole world. The "Doctrine of Tone-Sensations" is the speculation of a genius, the expression of an artistic intuition, which points the way - though it be only by the symbolism of a physical analogy, and as in a picture - along which further research will have to advance. We must therefore be careful not to throw overboard much that is valuable along with the deficiencies that have to be set aside. For what reasons Helmholtz himself took so little notice of criticism, I do not know. But in the last disposition which he made, according to which the text of the *Tone-Sensations* was to remain unaltered after his death, I think that he acted rightly.

22.

To a person accustomed to looking at things from the point of view of the theory of evolution, the high development of modern music and the spontaneous and sudden appearance of great musical talent seem, at first glance, a most singular and mysterious phenomenon. What can this remarkable development of the power of hearing have to do with the preservation of the species? Does it not far exceed the measure of the necessary or the merely useful? What can possibly be the significance of a fine discrimination of pitch? Of what use to us is the sense of intervals, or of the acoustic colorings of orchestral music?

As a matter of fact, the same question may be proposed with reference to every art, no matter from what province of sense its material is derived. The question is pertinent, also, with regard to the intelligence of a Newton, an Euler, or their like, which apparently far transcends the necessary measure. But the question is most obvious with reference to music, which satisfies no practical need and for the most part depicts nothing. Music, however, is closely allied to the decorative arts. In order to be able to see, a person must have the power of distinguishing the directions of lines. If he has a fine power of distinction, such a person may acquire, as a sort of collateral product of his education, a feeling for agreeable combinations of lines. The case is the same with the sense of color-harmony following upon the development of the power of distinguishing colors, and so, too, it undoubtedly is with respect to music.

We must bear in mind that what we call talent and genius, however gigantic their achievements may appear to us, constitute but a slight departure from normal endowment. Talent may be resolved into the possession of psychical power slightly above the average 'in a certain province. And as for genius, it is talent supplemented by a capacity of adaptation extending beyond the youthful period, and by the retention of freedom to overstep routine barriers. The naivety of the child delights us, and produces almost always the impression of genius. But this impression as a rule quickly disappears, and we perceive that the very same utterances which, as adults, we are wont to ascribe to freedom, have their source, in the child, in a lack of fixed character.

Talent and genius, as Weismann has aptly shown, [48] do not make their appearance slowly and by degrees in the course of generations; nor can they be the result of accumulated effort and practice on the part of ancestors; but they manifest themselves spontaneously and suddenly. Taken in connexion with what has been said above, this, too, is intelligible, if we will but reflect that descendants are not exact reproductions of their immediate ancestors, but exhibit the qualities both of their immediate and of their more distant ancestors and relatives with some variations, now slightly diminished, now slightly augmented in amount. The comparison of several children of the same two parents is very instructive on this point. To deny the influence of pedigree on psychical dispositions would be as unreasonable as to reduce everything to it, as is done, whether from narrow-mindedness or dishonesty, by modern fanatics on the question of race. Surely everyone knows from his own experience what rich psychical acquisitions he owes to his cultural environment, to the influence of long vanished generations, and to his contemporaries. The factors of development do not suddenly become inoperative in postembryonic life. [49]

[1] Apart from details, I have held the position here taken up since 1865. Stumpf, whom I must here thank for the repeated consideration of my work, has many points of detail (*Tonpsychologie,* Leipzig, 1883, Vol. I.) that appeal to me. The view expressed on page 119 of his work, however, seemed incompatible with the principle of parallelism, my fundamental axiom of research; though the remark which he directs against Lipps (Beitrdge zur Akuitik, Vol. I., p. 47, footnote), represents an approximation to my point of view. Compare my note, "Zur Analyse der Tonempfindungen," *Sitzungsberichte der Wiener Akademie,* Vol. XCII., II. Abth., p. 1282 (1895).

[2] Schopenhauer, *Die Welt als Wille und Vorstellung.*

[3] H. Berg, *Die Lust an der Musik*, Berlin, 1879.

[4] Fechner, notably, has emphasized the significance of association for aesthetics.

[5] Mach, "Bemerkungen über die Function der Ohrmuschel" (Tröltsch's *Archiv für Ohrenheilkunde,* New Series, Vol. III., p 72). - Compare also Mach and Fischer, "Die Reflexion und Brechurig des Schalles, *Pogg. Ann.,* Vol. CXLIX., p. 221; A. Steinhauser, *Theorie des binaurealen Hörens,* Vienna, 1877.

[6] I once had occasion to observe that tame marmots that were quite insensible to deep and loud noises, were suddenly frightened, and always rushed into hiding, whenever anyone produced a high-pitched noise by rubbing straw or crackling paper. Children a few months old are also very sensitive to such noises.

[7] Helmholtz, *Die Lehre von den Tonempfindungen,* 1st ed., Brunswick, 1863.

[8] Cp. "Zur Geschichte der Akustik" in my *Popular-wissenschaftliche Vorlesungen.*

[9] The pth harmonic of n coincides with the qth of m when $pn = qm$, that is $m = (p/q) n$, where p and q are whole numbers.

[10] I gave an account of part of my experiments, which were a continuation of Dvorak's researches on the after-images due to variations of stimulus (1870), in the August number of *Lotos,* 1873. I have never before mentioned the experiments relative to the excitement of pianotones by explosions. It will not be amiss, perhaps, if I do so here. Pfaundler, S. Exner, Auerbach, Brücke, W. Kohlrausch,

Abraham and Brühl, and others, subsequently treated the same questions in detail, and from various points of view.

[11] R. Konig, *Quelques expériences d'acoustique,* Paris, 1882.

[12] L. Hermann, "Zur Lehre von der Klangwahrnehmung," Pflüger's *Archiv,* Vol. LVI., 1894, p. 467.

[13] As long ago as 1867, I instituted experiments with a special kind of siren, very similar to one of König's apparatus. The casing of a cylinder was fitted with rings in which were cut wave-shaped slips in similar pairs capable of mutual displacement towards one another, so that the intensity and the phase of the partial tone under investigation could be varied at will. But it appeared on experiment that the wave-shaped slits did not yield any simple tones when air was blown against them through a slit parallel to the ordinate of the waves. As my apparatus was still pretty imperfect, and did not fulfil its purpose, which was to compound a sound from partial tones of given intensity and phase, I have not published any account of these experiments.

[14] König, *op. cit.* He got his tones by a very powerful tuning-fork, and I could not help conjecturing when his book appeared that, in connexion with his observations of the beats, the overtones came into play in various ways. Since then Stumpf has actually demonstrated the co-operation of such overtones (Wiedemann's *Annalen,* New Series, Vol. LVII., p. 660). Thus the theory of Helmholtz is safe on this side. Still, the objection remains that the objective compound tones do not exist (König, Hermann), and that subjective compound tones occur in circumstances which are not compatible with Helmholtz' theory (Hermann). Cp. also M. Meyer, "Zur theorie der Differenztöne und der Gehorsempfindungen Ueberhaupt," *Zeitschrift für Psychologie.* Vol. XVI., p. I.

[15] Hermann, "Zur Theorie der Kombinationstöne," Pflüger's *Archiv,* Vol. XLIX., 1891, p. 499.

[16] Hermann, Pflüger's *Archiv,* Vol. LVI., p. 493.

[17] *Harmoniesystem in dualer Entwicklung* (Dorpat, 1866), p. 30.

[18] [The lowest of the harmonics common to all I term the coincident or phonic harmonic. - Von Oettingen, *Harmoniesystem in dualer Entwicklung,* p. 32. *Quoted by translator.*]

[19] A popular statement of the principle of duality, of which Euler (*Tentamen novae theoriae musicae,* p. 103), D'Alembert (*Elémens de musique,* Lyons, 1766), and Hauptmann (*Die Natur der Harmonik und Metrik,* Leipzig, 1853, translation by W. E. Heathcote, London, Swan Sonnenschein & Co., 1888), had all a faint inkling, is to be found in my *Popular Scientific Lectures* (Chicago, 1894), under the caption "Symmetry" (originally published in 1872). Perfect symmetry, such as is found in the province of sight, cannot be imagined in music, since sensations of tone do not constitute a symmetrical system.

[20] I am here chiefly following Stumpf's *Beiträge zur Akustik und Musikwissenschaft,* Heft 1, Leipzig, 1898.

[21] Such explanations were attempted by Leibniz and Euler, and have been revived in more recent times by Oppel, later by Lipps (*Psychologische Studien,* 1885), and finally, in a number of voluminous works, by A. J. Polak (*Ueber Zeiteinheit in bezug auf Konsonanz, Harmonic und Tonalität,* Leipzig, 1900; *Ueber*

Tonrhythmik und Stimmführung, Leipzig, 1902; *Die Harmonisierung indischer, türkischer und japanischer Melodien,* Leipzig, 1905).

[22] C. Stumpf, "Geschichte des Konsonanzbegriffes," *Abhandlungen der Münchener Akademie,* 1897.

[23] C. Stumpf, *Beiträge zur Akustik,* Heft 1, p. 50.

[24] Mach, "Zur Theorie des Gehörorgans" (*Sitzungsberichte der Wiener Akademie,* 1863).

[25] Cp. my "Bemerkungen zur Lehre vom raumlichen Sehen" (*Fichte's Zeitschrift für Philosophie,* 1865), and see my *Popular Scientific Lectures.*

[26] Einleitung in die Helmholtzsche Musiktheorie, Graz, 1866. See the Preface and pp. 23 et seq., 46 and 48.

[27] The view that different end-organs respond to different rates of vibration is too well supported by the production of beats by neighboring tones, and by other facts adduced by Helmholtz, and is too valuable for the comprehension of the phenomena, to be again relinquished. The view here presented utilizes the facts disclosed, notably by Hering, in the analysis of color-sensations.

[28] *Zur Theorie des Gehörorgans,* 1863. By means of experiments carried out by me in co-operation with Kessel ("Ueber die Akkommodation des Ohres," *Sitzungsberichte der Wiener Akademie,* Vol. LXVI., part 3, October 1872), I obtained a proof of the variable disposition and capacity for resonance of the anterior auditory apparatus in the case of different tones; this was done by means of microscopic observation of sound-vibrations conducted through a tube. By introducing a tube, and making our observations by means of a microscopic ear-mirror constructed for the purpose, we tried to detect a similar spontaneous change of disposition in the living ear, but unsuccessfully. I have, however, subsequently been inclined to doubt whether the powerful vibrations which are observed in this way, would of themselves be decisive of the question, since, unless they were muted, they could scarcely penetrate into the labyrinth without doing damage. Consequently, as long as the vibrations cannot be observed with certainty in a normal living ear, it will scarcely be possible to decide this question definitely. A method of light-interference might lead to the desired result. But such a method would have to be of particularly simple form, if it is to be applicable under the difficult conditions of the living ear.

[29] Cp. my *Einleitung in die Helmholtzsche Musiktheorie,* p. 29.

[30] Cp. my *Grundlinien der Lehre von den Bewegungsempfindungen.* p. 58.

[31] E. Kulke, *Ueber die Umbildung der Melodie. Ein Beitrag zur Entwicklungslehre,* Prague, Calve, 1884.

[32] Cp. Fechner, *Ueber einige Verhältnisse des binocularen Sehens,* Leipzig, 1860, p. 536. I have myself often tried such experiments.

[33] Euler, *Tentamen novae theories musicae,* Petropoli, 1739, p. 36.

[34] I cannot understand how any one can still maintain the theory of the temporal coincidence of impulses. At one time I replaced A. Seebeck's experiment by what I believe to be a better procedure ("Ueber einige der physiologischen Akustik angehörige Erscheinungen," *Berichte der Wiener Akademie,* 26th June 1864), but I never could detect any periodicity in the nerve-process connected with sensation. At that time it was not known that beats are never observed between a subjective tone and an adjacent objective tone, or between subjective ones; but

the fact cannot now be doubted. Cp. Stumpf's interesting paper on "Beobachtungen über subjective Töne und über Doppelthören" (*Zeitschrift für Psychologie und Physiologie der Sinnesorgane,* Vol. XXI., pp. 100-121). The subjective tones that arise in my ear generally last too short a time for me to experiment with them. Still I did succeed not long ago (1906) in getting to the piano with a very clear and constant c-sharp, and convincing myself that when a c-sharp a shade deeper was lightly struck on the piano, no beats were demonstrable. For me, indeed, this demonstration was superfluous, for I hold the opposite supposition to be physiologically inadmissible. But Stumpf's observations on the consonance and dissonance of subjective tones without beats are very important.

[35] It is questionable therefore whether animals which have so small a measure of time that their voluntary movements produce a musical note hear in the ordinary sense, or whether with them it is not rather touch which makes on us the impression of hearing. Cp., for example, the admirable experiments and observations of V. Graber ("Die Chordotonalen Organe," *Arch. für Microskop. Anat.,* XX., p. 506). - Cp. also my *Bewegungsempfindungen,* p. 123. This conjecture has subsequently been confirmed in many ways. See *Populär-wissenschaftliche Vorlesungen,* 3rd ed., p. 401.

[36] As V. Hensen, for example, has observed.

[37] Thus, to take a very simple example, we might make $f(n)$ = k. log. n.

[38] The above exposition will be found in a rather conciser and slightly different form in my note "Zur Analyse der Tonempfindungen" (Sitzungsberichte der Wiener Akademie, December 1885), where I have tried to analyse sensations of tone on the analogy of sensations of color, of which the analysis has been carried very much further. Every vibration-number disengages a few specific energies in a ratio which depends on the vibration-number in question, The excitability of these energies is different at different points on the retina. Analogous relations are assumed mutatis imitandis for sensations of tone. It seemed originally that in both cases there must be an infinite variety of sensations corresponding to the infinite variety of the physical stimulus. But in both cases psychological analysis leads to the conclusion that a smaller number of sensations is to be assumed, and that, on the principle of parallelism, we have to think of these sensations, not as immediately dependent on the complicated physical stimulus, but as immediately dependent on the psycho-physical process which is as simple as they are themselves.

[39] C. Stumpf, *Beiträge zur Akustik und Musikwissenschaft,* Heft I, pp. 17, 18.

[40] Mach, "Zur Theorie des Gehörorgans," *Sitzungsberichte der Wiener Akademie, Vol.* LVIII., July 1863. Also Helmholtz, *Die Mechanik der Gehörknöchelchen,* 1869.

[41] Ewald, *Eine neue Hörtheorie,* Bonn, 1899.

[42] Ewald, Pflüger's *Archiv,* 1903, Vol. XCIII., p. 485.

[43] *Optisch-akustische Versuche,* Prague, 1872, p. 93,

[44] Friesach, *Berichte der Wiener Akademie,* 1867, Vol. LVL, 2nd part, p. 316.

[45] Stöhr, "Ueber Unterbrechungstöne," *Deutsche Revue,* July, 1904. Mach and Kessel some time ago pointed out the necessity of attacking the problems of acoustics from the point of view of asymmetry; see "Die Funktion der Trommelhöhle," *Berichte der Wiener Akademie,* Vol. LXVI., 3rd part, 1872; Stöhr,

"Klangfarbe oder Tonfarbe," *Süddeutsche Monalshefte,* Munich and Leipzig, July 1904. In this paper Stöhr is aiming, though by a different route, at a goal not far removed from my own.

[46] Hermann, Pflüger's *Archiv,* Vol. LVI., pp. 494 sqq., 1894.

[47] Perhaps this assumption would still remain valuable if it also succeeded in providing a basis for a satisfactory physical theory of resonance.

[48] Weismann, *Ueber die Vererbung,* Jena, 1883 (English translation, Clarendon Press, Oxford, 1889), p. 43.

[49] Cf. the sound and sober view of R. Wallaschek, Anfdnge der Tonkunst) Leipzig, 1903, pp. 291-298.

XIV. Influence of the Preceding Investigations on Our Conception of Physics

[1]

1.

WHAT gain does physics derive from the preceding investigations? In the first place, a very widespread prejudice is removed, and with it, a barrier. There is no rift between the psychical and the physical, no inside and outside, no "sensation" to which an external "thing," different from sensation, corresponds. There is but one kind of elements, out of which this supposed inside and outside are formed - elements which are themselves inside or outside, according to the aspect in which, for the time being, they are viewed. The world of sense belongs both to the physical and the psychical domain alike. As, in studying the behaviour of gases, by disregarding variations of temperature we reach Mariotte's law, but by expressly considering them, Gay Lussac's, while throughout our object of investigation remains the same; so, too, we are studying physics in its broadest signification when in searching into the connexions of the world of sense we leave our own body entirely out of account; whereas we are studying the psychology or physiology of the senses when we direct our main attention to the body and above all to our nervous system. Our body, like every other, is part of the world of sense; the boundary-line between the physical and the psychical is solely practical and conventional. If, for the higher purposes of science, we erase this dividing-line, and consider all connexions as equivalent, new paths of investigation cannot fail to be opened up.

We must regard it as an additional gain that the physicist is now no longer overawed by the traditional intellectual implements of physics. If ordinary "matter" must be regarded merely as a highly natural, unconsciously constructed mental symbol for a relatively stable complex of sensational elements, much more must this be the case with the artificial hypothetical atoms and molecules of physics and chemistry. The value of these implements for their special, limited purposes is not one whit destroyed. As before, they remain economical ways of symbolizing experience. But we have as little right to expect from them, as from the symbols of algebra, more than we have put into them, and certainly not more enlightenment and revelation than from experience itself. We are on our guard now, even in the province of physics, against overestimating the value of our symbols. Still less, therefore, will the monstrous idea of employing atoms to explain psychical processes ever get possession of us; seeing that atoms are but the symbols of those peculiar complexes of sensational elements which we meet with in the narrow domains of physics and chemistry.

2.

The fundamental views of mankind are formed by a natural process of adaptation to a narrower or wider sphere of experience and thought. It may be that the physicist is still satisfied with the notion of a rigid matter, of which the only changes are movements, or changes of place. Of such a thing as this the physiologist or psychologist can make nothing at all. But any one who has in mind the gathering up of the sciences into a single whole, has to look for a conception to which he can hold in every department of science. Now if we resolve the whole material world into elements which at the same time are also elements of the psychical world and, as such, are commonly called sensations; if, further, we regard it as the sole task of science to inquire into the connexion and combination of these elements, which are of the same nature in all departments, and into their mutual dependence on one another; we may then reasonably expect to build a unified monistic structure upon this conception, and thus to get rid of the distressing confusions of dualism. Indeed, it is by regarding matter as something absolutely stable and immutable that we actually destroy the connexion between physics and physiology.

Epistemological criticism can indeed do no one any harm, but the specialist, - the physicist, for instance, - has no reason to allow himself to be troubled overmuch by such speculations. Acuteness of observation and a felicitous instinct are very safe guides for him. His conceptions, in so far as they prove to be inadequate, will be best and most quickly corrected by the facts. But when it is a question of bringing into connexion two adjacent departments, each of which has been developed in its special and peculiar way, the connexion cannot be effected by means of the limited conceptions of a narrow special department. By means of more general considerations, conceptions have to be created which shall be adequate for the wider domain. Every physicist is not an epistemologist, nor ought every physicist to be one, even if it were possible. Special research demands a man's full energies; but so does epistemology.

Not long after the first edition of this book was published, I was lectured by a physicist on the misguided way in which I had conceived my task. In his opinion, it was impossible to analyse the sensations as long as the paths of the atoms in the brain were unknown; and when they were known everything else would follow of itself. Of course I had not much use for utterances such as these, which, had I been a young man of the period of Laplace, might have fallen upon fertile ground and have developed into a psychological theory based on "concealed movements." The effect which they had was, however, to make me offer a silent apology to Dubois-Reymond with his *Ignorabimus,* - a dictum which up to that moment I had regarded as the greatest mistake. After all, Dubois-Reymond's recognition of the insolubility of his problem was an immense step in advance; this recognition removed a weight from many men's minds, as is shown by the success of his work, a success which is otherwise scarcely intelligible. [2] He did not, indeed, take the further important step of seeing that the recognition of a problem as insoluble in principle, must depend on a mistaken way of stating the question. For he too, like countless others, took the instruments of a special science to be the actual world.

3.

The sciences may be distinguished according to the matter of which they treat, as also by their manner of treating it. But all science has for its aim the representation of facts in thought, either for practical ends, or for removing intellectual discomfort. Resuming the terminology of the "Introductory Remarks," science, we may say, arises when the combination of the other elements is imitated by the elements $\alpha \beta \gamma$... For example, physics (in its broadest signification) arises by the representative reproduction of the elements $A B C$ in their relations to one another; the physiology or psychology of the senses, through reproducing in like manner the relations of $A B C$...to $K L M$...; physiology, through reproducing the relations of $K L M$... to one another and to $A B C$...; while the reproducing of the $\alpha \beta \gamma$...themselves by other $\alpha \beta \gamma$...leads to the psychological sciences proper.

Now one might be of the opinion, say, with respect to physics, that the portrayal of the sense-given facts is of less importance than the atoms, forces, and laws which form, so to speak, the nucleus of the sense-given facts. But unbiassed reflexion discloses that every practical and intellectual need is satisfied the moment our thoughts have acquired the power to represent the facts of the senses completely. Such representation, consequently, is the end and aim of physics; while atoms, forces, and laws are merely means facilitating the representation. Their value extends as far, and as far only, as the help they afford.

4.

Our knowledge of a natural phenomenon, say of an earthquake, is as complete as possible when our thoughts so marshal before the eye of the mind all the relevant sense-given facts of the case that they may be regarded almost as a substitute for the phenomenon itself, and the facts appear to us as old familiar figures, having no power to occasion surprise. When, in imagination, we hear the subterranean thunders, feel the oscillation of the earth, figure to ourselves the sensation produced by the rising and sinking of the ground, the cracking of the walls, the falling of the plaster, the movement of the furniture and the pictures, the stopping of the clocks, the rattling and smashing of windows, the wrenching of the door-posts, the jamming of the doors; when we see in mind the oncoming undulation passing over a forest as lightly as a gust of wind over a field of grain, breaking the branches of the trees; when we see the town enveloped in a cloud of dust, hear the bells begin to ring in the towers; further, when the subterranean processes, which are at present unknown to us, shall stand out in full sensational reality before our eyes, so that we shall see the earthquake advancing as we see a waggon approaching in the distance till finally we hear the earth shaking beneath our feet, - then more insight than this we cannot have, and more we do not require. If we cannot combine the partial facts in their right proportions without the aid of certain auxiliary mathematical conceptions or geometrical constructions, it yet remains true that these constructions merely enable our thoughts to grasp gradually and piecemeal what they are unable to grasp all at once. But these auxiliary conceptions would be devoid of value, could we not reach, by their help, the graphic representation of the sense-given facts.

When I see in thought a white beam of light which falls upon a prism issue forth in a fan-shaped band of colors having certain angles which I can specify beforehand; when I see the real spectrum-image, obtained upon a screen by interposing a lens, with Fraunhofer's lines occurring in it at points determinable in advance; when I see, in my mind, how these lines alter their position on the prism being turned, on its substance being changed, or on the thermometer in contact with it altering its register, then I know all that I can require. All auxiliary conceptions, laws, and formulae, are but quantitative norms, regulating my sensory representation of the facts. The latter is the end, the former are the means.

5.

The adaptation of thoughts to facts, accordingly, is the aim of all scientific research. In this, science only deliberately and consciously pursues what in daily life goes on unnoticed and of its own accord. As soon as we become capable of self-observation, we find our thoughts, in large measure, already adjusted to the facts. Our thoughts marshal the elements before us in groups copying the order of the sense-given facts. But the limited supply of thoughts cannot keep pace with the constantly augmenting sweep of experience. Almost every new fact necessitates a new adaptation, which finds its expression in the operation of judgment.

This process is easily followed in children. A child, on its first visit from the town to the country, strays, for instance, into a large meadow, looks about, and says wonderingly: "We are in a ball. The world is a blue ball." [3] Here we have two judgments. What is the process accompanying their formation? In the first instance, the already existing sensational representation "we" (himself and his companions) is filled out into a single image by means of the representation of a ball, which also already existed. Similarly, in the second judgment, the image of the "world" (*i.e.,* all the objects of the environment) is supplemented by combination with the image of an enveloping blue ball (the representation of which must also have been present, since otherwise the name for it would have been wanting). A judgment is thus always a supplementing of a sensational presentation in order to represent more completely a sensational fact. If the judgment can be expressed in words, then the new presentation is never more than a combination of formerly established memory-images, which can also be elicited in a person addressed by words.

The process of judgment, therefore, in the present case, consists in the enrichment, extension, and supplementation of sensational presentations by other sensational presentations under the guidance of sense-given facts. If the process is over, and the image has assumed a familiar shape, making its appearance in consciousness as a completed presentation, then we have no longer to do with a judgment but merely with a simple memory. [4] It is to the formation of such intuitive knowledge, as Locke calls it, that natural science and mathematics mainly owe their growth. Consider, for example, the following statements: (1) the tree has a root; (2) the frog has no claws; (3) the caterpillar is transformed into a butterfly; (4) weak sulphuric acid dissolves zinc; (5) friction electrifies glass; (6) an electric current deflects a magnetic needle; (7) a cube has six surfaces, eight corners, twelve edges. The first statement contains a spatial extension of the presentation "tree"; the second a correction of a presentation too hastily generalized from habit; the third, fourth, fifth, and sixth contain temporal extensions of their respective presentations. The seventh proposition is an example of geometrical intuitive knowledge.

6.

Intuitive knowledge of the sort just described impresses itself upon the memory, and makes its appearance in the form of recollections which spontaneously supplement every fact presented by the senses. The various facts are not exactly alike. But the component parts of the sensational presentation which are common to different cases are emphasized, and so we reach a principle which holds a paramount place in memory - the principle of *broadest possible generalization* or *continuity*. On the other hand, if memory is to do justice to the complexity of facts, and be of real practical use, it must conform to the principle of *sufficient differentiation*. Even the animal is reminded, by soft, bright red and yellow fruits (seen without exertion on the tree), of their sweet taste, and by green hard fruits (which are seen with difficulty), of their sour taste. The insect-hunting monkey snatches at everything that buzzes and flies, but avoids the yellow and black fly, the wasp. Here we have expressed, distinctly enough, the combined effort for the greatest possible *generalization and continuity* and for *practically sufficient differentiation* of memory. And both ends are attained by the same means, the selection and emphasis of those particular elements of the sensational presentations which are determinative of the direction which the thought must pursue to suit the experience. The physicist proceeds in quite an analogous manner, when he says (generalizing): All transparent solids refract incident light towards the perpendicular, and when he adds (differentiating): amorphous bodies and isomeric crystals simply, the rest doubly.

7.

A great part of our mental adaptation takes place unconsciously and involuntarily, under the guidance of the facts presented to the senses. If this adaptation has become sufficiently comprehensive to embrace the vast majority of the occurring facts, and subsequently we come upon a fact which runs violently counter to the customary course of our thought without our being able to discover at once the determinative factor likely to lead to a new differentiation, then a problem arises. The new, unusual, and marvellous acts as a stimulus, which irresistibly attracts the attention. Practical considerations, or even bare intellectual discomfort, may engender the will to remove the contradiction or to adapt our thoughts to the new fact. Thus arises purposive thought-adaptation, or investigation.

For example, we have all, at some time or another, quite contrary to the common run of our experience, observed a lever or pulley lifting a large weight by means of a small one. We seek the differentiating factor, which cannot be immediately disclosed to us by the fact itself as given to the senses. It is only when, comparing various similar facts, we have noted the influence of the weights, and of the arms of the lever, and by our own exertions have reached the abstract concepts of "moment" or "work," that the problem is solved. "Moment" or "work" is the differentiating element. When it has become a habit of thought to pay attention to "moment" or "work," the problem no longer exists.

8.

What do we do when we abstract? What is an abstraction? What is a concept? Is there a sensational presentation-image corresponding to the concept? I cannot represent to myself a man in general. I can at most represent to myself a particular man, or perhaps one combining such accidental peculiarities of different men as are not exclusive of each other. A universal triangle, which is at once right-angled and equilateral, cannot be imagined. Further, the image thus rising into consciousness at the name of the concept, and accompanying the conceptual process, is not the concept. In fact, generally, words, being designations which from necessity must be used to describe many particular presentations, are far from corresponding completely to any concept. A child who has seen for the first time a black dog and heard it named, soon afterward calls a large and swiftly-running black beetle, "dog"; or a pig or a sheep, "dog." [5] Any similarity whatever reminding him of the presentation to which the name was first given naturally leads to the use of the same name. The point of similarity need not be at all the same in the suc-

184

cessive cases. It may reside, for instance, at one time, in the color, at another in the motion, at another in the form, at another in the external covering; and so on. Thus there is no question of a concept. Thus a child calls the feathers of a bird "hairs"; the horns of a cow "feelers"; a shaving-brush, its father's beard, and the down of a dandelion, without distinction, "shaving-brush"; and so on. [6] Most adults treat words in the same manner, only less noticeably so, because they have a larger vocabulary at their disposal. The illiterate man calls a rectangle a "square"; occasionally he also calls a cube a "square," because of its rectangular boundaries. The science of language, and a number of authenticated historical examples, show that even nations do not act differently. [7]

A concept is never simply a completed presentation. In using a word denoting a concept, there is nothing involved in the word but a simple impulse to perform some familiar sensory operation, as the result of which a definite sensational element (the mark of the concept) is obtained. For example, when I think of the concept "heptagon," I enumerate the angles of a figure visibly before me or of its image in my consciousness; and when in so doing I reach *seven,* in which case the sound, the numeral, or my finger may announce the sensational mark of the number, then by this very act the given presentation falls under the given concept. In speaking of a "square number," I seek to resolve the number given into components typified by the operation 5 x 5, 6 x 6, etc., the sensational characteristic of which, namely, the equality of the two factors multiplied, is patent. The same holds good of every concept. The activity excited by the word may be made up of a number of operations, one of which may contain the other. But the result is always a sensational element which was not present before.

In looking at or in imagining a heptagon, the fact of its having seven angles need not be present to my mind. This fact is distinctly cognized only on counting. Frequently, the new sensational element may be so obvious (as it is, for instance, in the case of the triangle) that the operation of counting seems unnecessary. Such cases, however, are exceptional, and constitute the main source of misunderstandings concerning the nature of concepts. In the case of conic sections (the ellipse, parabola, hyperbola) I do not directly *see* that these curves all fall under the same concept; but I can discover the fact by cutting a cone, and by constructing the equation for conies.

When, therefore, we apply abstract concepts to a fact, the fact merely acts upon us as an impulse to a sensational activity, which introduces new sensational elements, which in their turn may determine the subsequent course of our thought in harmony with the fact. By this activity we enrich and extend the fact, which before was too meagre for us. We do what the chemist does with his colorless solution of salts, when by a definite operation he elicits from it a yellow or brown precipitate, which has the power to differentiate the course of his thought. The concept of the physicist is a definite reaction-activity, which enriches a fact with new sensational elements.

A small measure of sensational endowment and a very low degree of mobility are sufficient for the formation of concepts, as is shown by the history of the mental development of the blind, deaf and dumb Laura Bridgman, which has been made generally accessible by Jerusalem's interesting little book. [8] In Laura Bridgman the sense of smell was almost entirely lacking; her only channel for the perception of disturbances and sound-vibrations was the soles of her feet and her finger-tips - her skin, in short; yet she succeeded in forming simple concepts. By walking about and by moving her hands she discovers the tactual signs (the class-characteristics) of a door, a chair, a knife, and so forth. The power of abstraction does not, indeed, go very far. The most abstract concepts to which she was able to attain seem to have been the numbers. On the whole her mental processes remained, naturally enough, attached to specific presentations. Evidence of this is afforded by her taking the sums in a school-book to be specially intended for her (*op. cit.*, p. 25), and her idea that heaven, or the world beyond, was a school (*op. cit.*, p. 30).

9.

To revert to an earlier example, when we see a lever, we are impelled to measure the length of its arms, to weigh its weights, and to multiply the numbers representing the lengths of its arms by the numbers representing the values of its weights. If the same sensational numerical symbol corresponds to both products, we expect equilibrium. We have thus gained a new sensational element which was not antecedently given in the bare fact itself, but which now differentiates the course of our thought. If we will keep well in mind the fact that conceptual thought is a reaction-activity which must be thoroughly practised, we shall understand the well-known fact that no one can familiarize himself with mathematics or physics or with any natural science by mere reading without practical exercise. Understanding here depends entirely on action. In fact, it is impossible in any province to grasp the higher abstractions without a practical working knowledge of its details.

Facts, then, are extended and enriched, and ultimately again simplified, by conceptual handling. For when the new determinative sensational element is found (say, the number representing the virtual moments of the lever), then the attention is directed to this alone, and the most diverse groups of facts are found to resemble and not to resemble each other solely in virtue of this element. Thus here also, as in the case of intuitive knowledge, everything is reducible to the discovery, selection, and emphasis of the determinative sensational elements. Investigation here only reaches by a roundabout way what is immediately presented to intuitive cognition.

The chemist with his re-agents, the physicist with his measuring-rod, scales, and galvanometer, and the mathematician all treat facts in precisely the same way; the only difference being that the latter needs to go least outside the elements $\alpha\,\beta\,\gamma...K\,L\,M$ in his extension of facts. The aids of the mathematician are always conveniently at hand. The investigator and all his thought are a fragment only of nature, like everything else. There is no real chasm between him and the other fragments. All elements are equivalent.

On the preceding theory, the essence of abstraction is not exhausted by terming it (with Kant) negative attention. It is true that, in abstracting, the attention is turned away from many sensational elements, but on the other hand, it is turned towards other new sensational elements; and precisely this latter fact is the essential feature. Every abstraction is founded on the prominence given to certain sensational elements.

10.

In the foregoing exposition of my views I have left what I wrote in 1886 unaltered; but at the same time I should like to refer the reader to the further explanations contained in a later work of mine. 1 In the second edition of the *Prinzipien der Wärmelehre* (1900), I have also mentioned the works of H. Gomperz and Ribot, which have appeared since 1897; these works contain investigations which in many respects have a certain affinity to my own. Both Gomperz and Ribot exclude scientific concepts from the scope of their inquiry, and treat only of such vulgar concepts as have been fixed in the words of the common speech of everyday intercourse. I, on the contrary, am of the opinion that the nature of concepts is necessarily much more clearly displayed in scientific concepts, which have been consciously formed and applied, than in vulgar concepts. The latter are so vague that they can scarcely be reckoned as proper concepts at all. The words of ordinary speech are simply familiar signs which occasion equally familiar habits of thought. The conceptual content of such words, in so far as it has any definite form at all, is scarcely present to consciousness, as Ribot also found by his statistical experiments. I have no doubt that, if Ribot and Gomperz had framed their inquiry so as to include scientific concepts also, my agreement with them would be even more far-reaching than it actually is.

We have chosen statical moment as a simple example of a concept. Complicated concepts will require a complicated system of reactions, drawing upon more or less large parts of the central nervous system, and helping to create a correspondingly complicated system of sensational elements characterizing the concept. Probably the difficulties pointed out by J. von Kries are not insuperable on this theory. [10] (Cf. Ch. IV.)

11.

The facts given by the senses, therefore, are alike the starting-point and the goal of all the mental adaptations of the physicist. The thoughts which follow the sense-given fact immediately are the most familiar, the strongest, and the most intuitive. Where we cannot at once follow a new fact, the strongest and most familiar thoughts press forward to mould it into a richer and more definite shape. This process is the source of all the hypotheses and speculations of science, which all find their justification in the mental adaptation that develops them and ultimately gives them birth. Thus we think of the planets as projectiles, we figure to ourselves an electric body as covered with a fluid that acts at a distance, we think of heat as a substance that passes from one body to another, until finally the new facts become as familiar and as intuitive as the older ones, which we have used as mental helps. Even where immediate intuition is out of the question, the thoughts of the physicist, by carefully observing the principle of continuity and of sufficient differentiation, become ordered in an economically assorted system of conceptual reactions, which lead, at least by the shortest path, to intuitive knowledge. All calculations, constructions, etc., are merely intermediate means, proceeding step by step, and always using sense-perception as a support, to the attainment of this kind of intuition in cases where it cannot be attained immediately.

12.

Let us now consider the results of mental adaptation. Thoughts can adapt themselves only to what is constant in the facts; it is only the mental reconstruction of constant elements that can yield advantage in point of economy. Herein is contained the ultimate ground of our effort for continuity in thought, that is, for the preservation of the greatest possible constancy, and in this way, too, the results of the adaptation are rendered intelligible. [11] Continuity, economy, and constancy mutually condition one another they are really only different aspects of one and the same property of all sound thinking.

13.

The unconditionally constant we term substance. I see a body upon turning my eyes in its direction. I can see it without touching it, I can touch it without seeing it. Although the actual appearance of the component elements of the complex is determined in this way by certain conditions, I yet have these conditions too absolutely in my power to appreciate or notice them markedly. I regard the body, or the complex of elements, or the nucleus of this complex, as always present, whether, for the moment, it is the object of my senses or not. Having always ready the thought of this complex, or, symbolically, the thought of its nucleus, I gain the advantage of being able to predict, and avoid the disadvantage of being surprised. My behaviour is the same with regard to the chemical elements, which also appear to me unconditionally constant. Although here my mere willing it is not sufficient to make of the complexes in question sensational facts, and although in the present case external aids (for instance, bodies exterior to my own body) also are necessary, I yet leave these aids out of account as soon as they have become familiar to me, and look upon the chemical elements as simply constant. The man who believes in atoms treats them in an analogous way.

In the same manner as with the complex of elements corresponding to a body, we may also proceed, on a higher plane of thought-adaptation, with entire provinces of facts. In speaking of electricity, magnetism, light, and

188

heat, even when not associating special substances with these names, we yet ascribe constancy to these provinces of facts, leaving entirely out of account the familiar conditions under which they appear; and we hold the ideas which reproduce them always in readiness, thereby gaining an advantage similar to that explained above. When I say a body is "electric," far more memories arise in my mind, and my expectations are associated with far more definite groups of facts, than if I had emphasized, for instance, the attractions displayed in the single cases. Yet this hypostasizing may have its disadvantages, also. In the first place, in proceeding thus, we always follow the same historical paths. It may be important, however, to recognize that there is no such thing as a specific electrical fact, that every such fact can just as well be regarded, for example, as a chemical one, or as a thermal one, or rather that all physical facts are made up, ultimately, of the same sensational elements (colors, pressures, spaces, times), and that we are merely reminded by the term "electric" of that particular form in which we first became acquainted with the fact.

If we have once accustomed ourselves to regard the body, to and from which we can, at pleasure, turn our glance or our hand, as constant, then it is easy for us to do the same in cases in which the conditions of sensational manifestation lie entirely beyond our power - for example, in the case of the sun and moon, which we cannot touch, or of parts of the world which we have seen but once and shall perhaps never see again, or that we know only by description. Such a method of procedure may have high importance for an undisturbed and economical conception of the world, but it is certainly not the only legitimate method. It would be merely a consistent additional step, if we were to regard the whole past, which is, indeed, still present in its vestiges (since, for instance, we see the stars where they were thousands of years ago), and the whole future, which is present in germ (since, for example, our solar system will be seen where it now is, thousands of years hence), as constant. The entire passage of time, in fact, is dependent solely on conditions of our sensibility. Were a special purpose given, even this step might be hazarded.

14.

Really unconditioned constancy does not exist, as will be evident from the preceding considerations. We attain to the idea of absolute constancy only as we overlook or underrate conditions, or as we regard them as always given, or as we deliberately disregard them. There is only one sort of constancy, which embraces all the cases that occur, namely, constancy of connexion or of relation. Substance, again, or matter, is not anything unconditionally constant. What we call matter is a combination of the elements or sensations according to certain laws. The sensations connected with the different sense-organs of a particular man are dependent on one another according to laws, as are the sensations of different men. It is in this that matter consists. The older generation, especially the physicists and chemists, will be alarmed by this proposal not to treat matter as something absolutely constant, but to take as constant, instead, a fixed law of connexion among elements which in themselves seem extremely unstable. Even younger minds may find this conception difficult; but the view is inevitable, though I myself at one time went through a great struggle in order to arrive at it. We shall have to make up our minds to some such radical change in the method of our thought, if we want to escape the alternative of perpetually recurring helplessness in the face of these questions.

There can be no question of abolishing from ordinary everyday use the vulgar conception of matter which has been instinctively developed for this purpose. Moreover, all our concepts of physical measurement can be maintained, only receiving such critical elucidation as I have tried to carry out for mechanics, heat, electricity, etc. Purely empirical concepts here take the place of metaphysical. But science suffers no loss when a "matter," which is a rigid, sterile, constant, unknown Something, is replaced by a constant law, of which the details are still capable of further explanation by means of physico-physiological research. In doing this our object is not to create a new philosophy or metaphysics, but to promote the efforts, which the positive sciences are at this moment making, towards mutual accommodation.

15.

The propositions of natural science express only such constancies of connexion as: "The tadpole turns into a frog. Chlorate of sodium makes its appearance in the form of cubes. Rays of light are rectilinear. Bodies fall with an acceleration of 9.81 (m/sec$_2$)." When these constancies are expressed in concepts, we call them laws. Force (in the mechanical sense) is likewise merely a constancy of connexion. When I say that a body A exerts a force on a body B, I mean that B, on coming into contraposition with A, is immediately affected by a certain acceleration with respect to A.

The singular illusion, that the substance A is the absolutely constant vehicle of a force which takes effect immediately on B's being contraposed to A, is easily removed. If we, or more exactly speaking, our sense-organs, be put in the place of B, here a condition intervenes, which, seeing that it is possible at any time to fulfil it, is invariably disregarded, and thus A appears to us absolutely constant. Similarly, a magnet, which we see as often as we care to look in its direction, appears to us the constant vehicle of a magnetic force, which becomes operative only upon being brought near to a particle of iron, which we cannot, without noticing the fact, disregard as easily as we can ourselves. [12] The phrases, "No matter without force, no force without matter," which are but abortive attempts to remove a self-incurred contradiction, become superfluous when we recognize only constancies of connexion.

16.

Given a sufficient constancy of environment, there is developed a corresponding constancy of thought. By virtue of this constancy our thoughts are spontaneously impelled to complete the half-observed facts. This impulse towards completion is not prompted by the individual facts as observed at the time; nor is it intentionally created; but we find it operative in ourselves entirely without our personal intervention. It confronts us like a power from without, yet as a power which continually accompanies and assists us, as a thing of which we stand in need, in order to supplement the facts. Although it is developed by experience, it contains more than is contained in the single experience. The impulse in a certain measure enriches the single fact. Through it the fact is more to us. With this impulse we have always a larger portion of nature in our field of vision, than the inexperienced man has with the single fact alone. For the human being, with his thoughts and his impulses, is himself merely a piece of nature, which is added to the single fact. This impulse, however, can lay no claim to infallibility, and there exists no necessity compelling the facts to correspond to it. Our confidence in it rests entirely upon the supposition, which has been substantiated by numerous trials, of the sufficiency of our mental adaptation, - a supposition, however, which must be prepared to be contradicted at any moment.

Not all our ideas representing facts have the same constancy. Whenever we have a special interest in the representation of facts, we endeavour to support and corroborate ideas of lesser constancy by ideas of greater constancy, or to replace them by the latter. Thus Newton conceived the planets as projectiles, although Kepler's laws were already well known, and the tides as attracted by the moon, although the facts of their movement had long been ascertained. We do not think that we understand the suction of a pump, or the flowing of a siphon, until we have mentally added the pressure of the air as holding the chain of particles together. Similarly we seek to conceive electrical, optical, and thermal processes as mechanical processes. This need of the support of weaker thoughts by stronger thoughts is also called the need of causality, and is the moving spring of all scientific explanations. We naturally prefer, as the foundation of this process, the strongest and most thoroughly tested thoughts, and these are given us by our much exercised mechanical functions, which we may test anew at any moment without many or cumbersome appliances. Hence the authority of mechanical explanations, especially those by pressure and impact. A corresponding and still higher authority attaches to mathematical thoughts, for in their development we stand in need of no extraneous means whatever, but, on the contrary, invariably carry most of the material for experimenting about with us. But if we are once apprised of this, the need of mechanical explanations is appreciably weakened. [13]

I have already often pointed out that a so-called "causal" explanation, also, is nothing more than the statement or description of an actual fact or of a connexion between facts, and I might here simply refer to the detailed discussions in my *Theory of Heat* and my *Popular Lectures*. But, as people who have not made a special study of physics always believe that they broaden the basis and increase the profundity of their thought if they assume a fundamental difference between a scientific description (for instance, of the development of an embryo) and a physical explanation, I may perhaps be allowed to add a few more words on the subject. When we describe the growth of a plant, we notice that there comes into play such an immense variety of circumstances varying from one case to another, that it is only in the broader features at most that our description can hope to apply universally, while as regards the minuter details it can only be accurate for the individual case. This is exactly what happens in physics when the circumstances are at all complicated; the only difference being that the circumstances are generally simpler and better known. That is why it is easier for us in physics to separate out the circumstances experimentally, and intellectually too, by means of abstraction. Schematization is easier. For the astronomers of antiquity, to describe the motion of the planets was a task analogous to what the description of a plant's development is for a modern botanist. The discovery of Kepler's laws depends upon a fortunate and fairly crude schematization. The more closely we consider a planet, the more individual does its movement become, and the less exactly does it follow Kepler's laws. Speaking strictly, all the planets move differently, and the same planet moves differently at different times. Now, when Newton gives a "causal explanation" of the planetary motions by showing that one particle of mass m acquires through another particle m' the acceleration $\varphi = \dfrac{km'}{r^2}$, and that the accelerations determined in the first particle by different particles are summed geometrically, he is only pointing out or describing facts, which, although by a roundabout path, yet have been reached by observation. Let us consider what the process is. The circumstances determining the planetary motions are first of all isolated from one another, that is to say, the individual particles of mass and their distances from one another. The relation between two particles of mass is very simple, and we think that we know all the circumstances, mass and distance, that determine this relation. If we have a description which has been found to be correct for a few cases, we extend it beyond the limits of experience and assume it to be universally correct, at the same time disregarding the possibility of any disturbance from an unknown and alien cause; in this, indeed, we may be mistaken, as we should be, for instance, if gravity were to turn out to be transferred through a medium and to require time for its transference. The modification of the relation is equally simple, as was pointed out, when to two particles a third is added, and to these a fourth, and so on. Thus Newton's description is not, in fact, the description of an individual case; it is a description in terms of elements. Newton, in describing the way in which the elements of mass are related to one another in the elements of time, indicates to us how we can describe in terms of the elements any individual case we like, according to his pattern. It is just the same with all the other cases that theoretical physics has mastered. But this does not mean that the essence of the description is changed in any way. What we have to

192

do with, is a general description in terms of elements. If we may remain satisfied with a representation of the phenomena by means of differential equations, a view which I long ago recommended (*Mechanik,* 1883, 4th ed., 1901, p. 530), and which seems to be coming more and more into favor, this actually amounts to the recognition that explanation is nothing more than a description in terms of elements. Every particular case can then be put together out of spatial and temporal elements, the relations between which are described by equations.

17.

It was said above that man himself is a fragment of nature. Let me illustrate this by an example. For the chemist a substance may be sufficiently characterized merely by his sensations. In this case the chemist himself supplies, by inner means, the whole wealth of fact necessary to the determination of his course of thought. But, in other cases, recourse to reaction by the help of external means may be necessary. When an electric current flows round a magnetic needle situated in its plane, the north pole of the needle is deflected to my left, if I imagine myself as Ampère's swimmer in the current. I enrich the fact (current and needle) which is insufficient in itself to define the direction of my thought, by introducing myself into the experiment by an inner reaction. I may likewise lay my watch in the plane of the circuit, so that the hand moves in the direction of the current. Then the south pole falls in front of, the north pole behind the dial. Or I may make of the circuit traversed by the current a sundial (on the plan of which the watch in fact was modelled [14]), so arranging it that the shadow follows the current. In this case the north pole will move towards the shadowed side of the plane of the current. The two last-mentioned reactions are *outward* reactions. The two species of reactions could not be made use of indiscriminately if a chasm existed between myself and the world. Nature is a single whole. The fact that the two species of reaction are not known in all cases, and that frequently the observer appears to be entirely without influence, proves nothing against the view advanced.

Right and left appear to us to be similar, in contrast to before and behind, and to above and below. Yet it is certain that they are only different sensations, overwhelmed by stronger similar sensations. The space of sensation thus has three strongly marked and essentially different directions. From a metrical point of view all directions of geometrical space are identical. Our immediate sensation represents symmetrical shapes as equivalent; but in physical respects they are by no means equivalent. Physical space also has three essentially different directions, which are most clearly manifested in a triclinic medium, in the behaviour of an electro-magnetic element. The same physical properties appear also in our own body, which is the reason why our bodies can be used as reagents in physical problems. If we had an exact physiological knowledge of an element of our bodies, we should thereby have laid, in all essentials, the foundation of our understanding of the physical universe. (Cp. Ch. VI.)

18.

I have repeatedly emphasized the unity of the physical and the psychical, but it is worthwhile to consider this unity once more in one of its special aspects. Our psychical life, in so far as we mean by that term our presentations, seems to be perfectly independent of physical processes; it seems to be a world in itself, with freer laws of its own, laws that are of a different order. But it is certain that this is a mere illusion, caused by the fact that only a very minute part of the traces of physical processes ever comes to life in our presentations. The circumstances determining this fragment are much too complicated to grasp, so that it is impossible to lay down any precise rule for its occurrence. In order to determine what thoughts a physicist, for instance, will connect with the observation of a particular optical fact, we should have to know the previous events of his life, the force of the impressions which they have left behind them, and the facts of the development of general and technical culture by which he has been influenced; and finally we should have to be in a position to take into account his mental disposition at the moment. To do all this, it would be necessary to enlist as an auxiliary the whole of physics, in its widest sense, and at an unattainably high stage of development. [15]

Let us now consider the other side of the picture. A physical fact, which we experience for the first time, is strange to us. If it happened in some quite different way from that in which it actually happens, it would not thereby be any more puzzling. The way in which it occurs appears to us not to be determined by anything, least of all to be uniquely determined. What it is that invests the way in which a physical fact occurs with the character of determinateness, can only be understood from our psychical development. The presentational part of our mental life is the agent which first draws the fact forth from its isolation, brings it into contact with an abundance of other facts, and then invests it with determinateness, in virtue of the necessity for agreement with those other facts and for he exclusion of contradiction. The science of psychology is auxiliary to physics. The two mutually support one another, and it is only when they are united that a complete science is formed. From our standpoint, the antithesis of subject and object, in the ordinary sense, does not exist. The question as to the greater or less degree of precision with which presentations copy the facts, is, like every other question, a problem of natural science.

19.

Whenever it happens, in a complexus of elements, that some of the elements are replaced by others, then a constancy of connexion of one kind becomes a different kind. In such cases it is desirable to discover a constancy which survives this change. J. R. Mayer was the first to feel this need, and satisfied it by enunciating his concept of "force," which corresponds to the technical mechanical concept of "work" (Poncelet), or more exactly to the more general concept of "energy" (Young). Mayer conceives this force (or energy) as something absolutely constant (as a store of something, as a material), thus harking back to the most stubborn intuitive notions. We perceive, from Mayer's struggle with expressions, and with general philosophical phrases (noticeable in the first and second of his treatises), that he at first felt instinctively and intuitively the urgent need of such a concept. But his great achievement was accomplished only by his adapting the existing physical concepts to the requirements of the facts as well as to his needs. [16]

20.

When the adaptation is adequate, the facts are spontaneously reproduced by the thoughts, and incompletely given facts are completed. Physics can act only as a quantitative norm regulating and giving a more precise conformation to the spontaneously flowing thoughts, suitably to practical or scientific needs. When I see a body thrown horizontally, the intuitive picture of a projectile in motion may rise before my mind. But the artilleryman or the physicist requires more. He must know, for example, that if on applying the measuring-rod M to the horizontal abscissae of the projectile's path, he can count to 1, 2, 3, 4..., he must, on applying the measure M' to the vertical ordinates, also count to 1, 4, 9, 16...in order to reach a point of the path. The function of physics consists, therefore, in teaching that a fact which, on a definite reaction R yields a sensory mark E, also yields, on the giving of a different reaction R', a second sensory mark E'. By this means it is possible to supply more exactly the deficiencies of incompletely given facts.

The introduction into physics of the universally comparable, or so-called "absolute" measurements, - the reduction of all physical measurements to such units as the centimetre, the gramme, and the second (length, mass and time), - has one peculiar result. There exists in any case a tendency to regard anything that can be physically grasped and measured, anything that can be stated in such a way as to become common property, as "objective" and "real," in contrast to the subjective sensations; and the absolute measures appear to give some support to this opinion, and to supply it with a psychological, if not with a logical, motive. It looks as if what we call "sensations" in the familiar sense, were something quite superfluous in physics. Indeed, if we look closer, the system of units of measurement can be still further simplified. For the numerical measurement of mass is given by a ratio of accelerations, and measurement of time can be reduced to measurement of angles or lengths of arcs. Consequently measurement of lengths is the foundation of all measurements. But we do not measure mere space; we require a material standard of measurement, and with this the whole system of manifold sensations is brought back again. It is only intuitional sense-presentations that can lead to the formulation of the equations of physics, and it is precisely in such presentations that the interpretation of these equations consists. Thus, though the equations only contain spatial numerical measurements, these measurements, also, are merely the ordering principle which tells us out of what members of the series of sensational elements we have to construct our picture of the world.

21.

I have elsewhere [17] shown that quantitative enunciations are only distinguished from qualitative by the fact that the former have reference to a continuum of homogeneous cases. On this view, the advantageous employment of equations for purposes of description would only be possible within a very limited field. There is, however, some prospect of enlarging this field by successive steps without any limit. This would be done in the following way. All possible optical sensations, though they cannot be measured, can be characterized and catalogued by means of numbers on psycho-physical methods. Thus any optical experience can be described by representing, by means of equations, the values of these numerical characteristics as dependent on the spatial and temporal co-ordinates, and on one another. And we shall have to hold that a result the same in principle can be obtained in the fields of the other senses also. Thus it is possible to assign a perfectly precise meaning to the expression used in Ch. II., above.

22.

The ascertainment of the dependence of the elements A B C on one another, K L M being disregarded, is the task of natural science, or of physics in its broadest sense. But, in reality, the A B C's are always also dependent on K L M. There are always equations of the form $f(A\ B\ C...K\ L\ M...) = 0$. Now since many different observers K L $M...$, K' L' $M'...$, K'' L'' $M''...$are involved, we succeed in eliminating the accidental influence of the variation of K L M, etc., and we thus obtain only the element that can be stated as common property, namely the pure dependence of the A B C's on one another. In this process the K L $M...$, K' L' $M'...$, are treated like physical instruments, each with its peculiarities, its special constants, and so forth, from which the results, as finally indicated, have to be set free. But if it is a question merely of the temporal connexion of one quantitative reaction with other quantitative reactions, as in the above dynamical example, the matter is then still simpler. Everything then turns on the ascertainment of equality or identity of the A B C's under like circumstances, - that is to say, under like K L M's, - which comes to saying that everything turns merely on the ascertainment of spatial identities. The kind of quality of the sensations is now indifferent; it is their equality that is alone decisive. And now a single individual suffices to fix relations of dependence which are valid for all individuals. Thus from this point onwards we have obtained a safe basis for the whole field of scientific research, - a fact which inures to the advantage of psychophysiology as well.

23.

The space of the geometrician is by no means merely the system of space-sensations (the senses of sight and touch), but consists rather of a body of conceptually idealized and formulated physical experiences, having the space-sensations as their point of departure. In the very fact of the geometrician's regarding his space as being of the same nature at all points and in all directions, he goes far beyond the space given to sight and touch, which by no means possesses this simple property (pp. 168, 181, sqq.). Without experience in physics the geometrician would never have reached this conception. The fundamental propositions of geometry have, as a fact, been acquired wholly by means of physical experiences, by the superposition of measures of length and of angles, by the application of rigid bodies to one another. Without propositions of congruence, no geometry. Apart from the fact that spatial images would not be produced in us without physical experience, we should, even granting their existence, never be able to apply them to one another and to test their congruence. When we feel compelled to imagine an isosceles triangle as having equal angles at its base, our compulsion is due to the remembrance of powerful past experiences. If the proposition had its source in "pure intuition," there would be no necessity for learning it. That discoveries may be made by sheer power of geometrical imagination, and are so made daily, merely proves that the memory of a given experience can reveal to the mind features which in the original observation escaped unnoticed; just as in the after-image of a bright lamp, new and previously unseen details may be discovered. Even the theory of numbers must be looked at in some such manner; its fundamental propositions can hardly be viewed as entirely independent of physical experience.

The cogency of geometry (and of all mathematics) is due, not to the fact that its theories are arrived at by some peculiar kind of knowledge, but only to the fact that its empirical material, which is particularly convenient and handy, has been put to the test very often, and can be put to the test again at any moment. Moreover, the province of space-experience is far more limited than that of the whole of experience. The conviction of having in all essentials exhausted this limited province soon arises and produces the necessary self-confidence. [18]

24.

A self-confidence similar to that of the geometrician is doubtless also possessed by the composer and the decorative painter, who have both gained, the former in the domain of sensations of tone, the latter in that of sensations of color, a broad and rich experience. To the one no space-figure will occur the elements of which are not well known to him, and the two others will meet with no new combinations of tone or of color that are unfamiliar to them. But the inexperienced beginner in geometry will be no less surprised and disappointed by the results of his activity than the young musician or decorator.

The mathematician, the composer, the decorator, and the student of natural science, when indulging in speculation, pursue quite analogous modes of procedure, despite the differences of their materials and aims. The mathematician, it is true, owing to his more limited material, has the advantage of the others as regards the certainty of his procedure; while the latter for the opposite reason is at a disadvantage as compared with the others.

25.

The distinction between physiological and geometrical space has proved to be unavoidable. But while geometrical insight is obtained by the spatial comparison of bodies with one another, time also cannot be left out of consideration, since it is impossible, in making such comparisons, to disregard the translation of bodies. Space and time stand in intimate connexion, thereby showing themselves to be relatively independent of the other physical elements. This is expressed in the fact that when bodies move their other properties remain relatively constant. It is precisely owing to this fact that pure geometry and mechanics are possible.

Space and time, closely considered, stand, as regards physiology, for special kinds of sensations; but, as regards physics, they stand for functional dependencies upon one another of the elements characterized by the sensations. When the spatial and temporal physiological indices, conditioned by the parts and processes of our body, are compared with one another in like physiological circumstances, we obtain relations of dependence of the physical elements on one another, that is, dependence of the elements of one body on those of another, and dependence of the elements of one process on those of another. On the basis of this result we can take the temporal and spatial determinations in a purely physical sense. Whatever coincides with the smaller part of a process which takes place continuously in one direction, is earlier in time. In a homogeneously filled space the position B is nearer than another position to the position A, when B is reached by the process starting from A earlier than the other position is reached. The straight line is the class-concept of the positions uniquely determined by the physical relations between two points, or infinitely small bodies. The position C is situated at the point of bisection of the straight line AB, when, in homogeneous space, processes starting from A and B reach that position in equal times, and reach it in a shorter time than any other position with which it shares the first property.

26.

The time of the physicist does not coincide with the system of time-sensations. When the physicist wishes to determine a period of time, he applies, as his standards of measurement, identical processes or processes assumed to be identical, such as vibrations of a pendulum, the rotations of the earth, etc. The fact connected with the time-sensation is in this manner made the subject of a reaction, and the result of this reaction, the number which is obtained, serves, in place of the time-sensation, to determine more exactly the subsequent movement of the thought. In like manner, we regulate our thoughts concerning thermal processes not according to the sensation of warmth which bodies yield us, but according to the much more definite sensation which is obtained from thermometrical reactions by simply noting the height of the mercury. Usually a space-sensation (a rotation-angle of the earth, or the path of the hand on the dial of a clock) is substituted for the sensation of time, and for this, again, a number is put. For example, if we represent the excess of the temperature of a cooling body over that of its surroundings by $\vartheta = \Theta e^{-kt}$, then t is this number.

The relation in which the quantities of an equation stand, is usually (analytically) a more general one than that which is meant to be represented by the equation. Thus in the equation $(x/a)^2 + (y/b)^2 = 1$ all possible values of x have an analytical meaning, and yield corresponding values of y. But if the equation be used to represent an ellipse, then only the values of $x < a$ and $y < b$ have a geometrical (or real) significance. Similarly, it would have to be expressly added, if this were not obvious, that the equation $\vartheta = \Theta e^{-kt}$ represents the real process only for increasing values of t.

If we imagine the natural course of different events, say the cooling of one body and the free descent of a second, represented by equations involving time, then time may be eliminated from these equations, and we may express, for example, the excess of temperature by means of the space traversed by the falling body. Thus viewed, the elements appear simply as dependent on one another. But the meaning of such an equation would have to be more exactly denned by adding that only increasing distances of descent or decreasing temperatures are to be inserted successively therein.

When we thus think of excess of temperature as determined by the space traversed by a falling body, the dependence is not an immediate one. On this point I agree with Petzoldt. [19] But the dependence is no more immediate when we assume excess of temperature to be determined by the angle of rotation of the earth. For no one will believe that the same temperature-values would continue to correspond to the same angular values, if the earth were to alter its velocity of rotation in consequence of some shock. But it seems to me to follow from such considerations that our postulates are merely provisional, and depend upon partial ignorance of the decisive part played by certain independent variables which are inaccessible to us. It is only in this sense that I would wish the reference which I once made to a certain absence of determination, to be understood. [20] This view, moreover, is perfectly compatible with the postulation of unique determinations, since such postulates are always laid down on the assumption of given circumstances, and with abstraction from unusual and unexpected changes. This way of looking at the matter is, as it seems to me, inevitable, when we reflect that the distinction emphasized by Petzoldt between simultaneous and successive dependencies, holds for intuitional presentation, but not for the equations

199

which are the norms regulating presentation quantitatively. The equations can only be of one kind, and can only express simultaneous dependencies. Towards indeterminism in the ordinary sense, the assumption, for instance, of freedom of the will in the sense used by many philosophers and theologians, - I have not the slightest inclination.

Time is not reversible. A warm body set in cool surroundings merely cools, and does not grow warm again. With larger, or later, time-sensations only smaller decreasing excesses of temperature are connected. A house in flames burns down but never builds itself up again. A plant does not decrease in size and creep into the earth, but grows out of it, increasing in size. The irreversibility of time reduces itself to the fact that the alterations in the values of physical quantities always take place in definite directions. Of the two analytical possibilities one only is actual. We do not need to see in this fact a metaphysical problem.

Changes can only be determined by differences. Where there are no distinctions there is no determination. The supervening change may increase the distinctions or it may diminish them. But if the differences had a tendency to increase, change would go on endlessly and aimlessly. The only assumption compatible with a general representation of the universe, or rather with the representation of our own limited environment, is that of a tendency, on the whole, to a diminution of differences. But if circumstances that set up differences did not make themselves felt by forcing their way into our environment, a time would soon come when nothing more would happen at all.

Again, we can conclude, with Petzoldt, from our own existence and from our bodily and spiritual stability, to the stability and to the uniqueness, as regards determination and direction, of the processes of nature. For not only are we ourselves a fragment of nature (see earlier this chapter), but it is the presence of these very properties in our environment that determines our existence and our thought (see *Popular Scientific Lectures,* 3rd edition, p. 250). But it will not do to build too confidently on this foundation, for organisms are peculiar fragments of nature, of very limited and moderate stability, which in point of fact are liable to destruction, and for the preservation of which a proportionately moderate amount of stability in the environment is sufficient. The most convenient course will therefore be to recognize the limits which are everywhere manifestly set to our knowledge, and to regard the effort towards unique determination as an ideal, which, so far as may be, we actualize in our thought.

I do not, of course, regard the statements which I wrote down at the time of my greatest intellectual ferment (1871), as secure against all attack, particularly as regards their form; nor do I by any means consider that Petzoldt's objections are dictated by a spirit of captiousness. I hope, however, that when I deal with the subject again at greater length, for I have only been able to touch upon it briefly here, I shall be able to bring about a full understanding, without at the same time giving up any essential part of my view. 1

[1] I have partly discussed the questions considered in this chapter, before (see my *History and Root of the Principle of the Conservation of Energy*, translated by P. E. B. Jourdain, Chicago, Open Court Publishing Co., 1911, and also the essay on "The Economical Nature of Physical Inquiry," first published in 1882, and now in my *Popular Scientific Lectures*, Chicago. 1894, and see my *Mechanik* and *Wärmelehre*). With regard to the idea of concepts as labor-saving instruments, the late Prof. W. James directed in conversation my attention to points of agreement between my writings and his essay on "The Sentiment of Rationality" (*Mind*, Vol. IV., p. 317, July 1879). This essay, written with refreshing vigor and impartiality, will be perused by everyone with pleasure and profit.

[2] Dubois-Reymond, *Ueber die Grenzen des Naturerkennens*, 1872, 4th. ed.

[3] This example is not fictitious, but was observed in the case of my three-year-old child. In this case what is actually attested is a physiological fact, but this has only been recognized at a late stage. Scientific astronomy begins in antiquity with similar naïf assertions, which it thinks are physical.

[4] I cannot here enter upon an examination of the process of judgment as such. But among recent works on the subject I should like to draw special attention to W. Jerusalem's *Die Urteilsfunktion* (Vienna, 1895). Though my own position is not that of the author's, I nevertheless have been greatly stimulated and instructed by many of the investigations of special points contained in his book. The physiological aspects, and in particular the biological function of judgment, are set forth in a very lively way. His conception of the subject in judgment as a centre of force can scarcely be called felicitous. On the other hand, it will certainly be readily admitted that in the early stages of culture and of the formation of language anthropomorphic conceptions exerted much influence. Other questions of a different kind are discussed by A. Stöhr in *Theorie der Namen* (1889), *Die Vieldeutigkeit des Urteils* (1895) and *Algebra der Grammatik* (1898). Of these works, those concerned with the relation between logic and grammar seem to me the most interesting.

[5] Thus the Marcomanni called the lions sent across the Danube by the Romans "dogs," and the Ionians called the χάμψαι of the Nile "crocodiles" from their likeness to the lizards (κροκόδειλοι) which, in Ionia, live in the walls. (Herodotos, II., 69.)

[6] All these examples are taken from actual experience.

[7] See W. D. Whitney, *Life and Growth of Language*, 1875.

[8] W. Jerusalem, *Laura Bridgman*, Vienna, Pichler, 1891. Cp. also L. W. Stern, *Helen Keller*, Berlin, 1905; Jerusalem, "Marie Heurtin," *Oesterreichische Rundschau*, Vol. III., pp. 292, 426 (1905).

[9] *Prinzipien der Wärmelehre*, 1896; 2nd ed., 1900, pp. 415, 422.

[10] J. von Kries, *Die materiellen Grundlagen der Bewusstseinserscheinungen*, Freiburg im Breisgau, 1898.

[11] Cp. my *The Science of Mechanics*, translated by T. J. M'Cormack, Chicago, Open Court Publishing Co., 1893, p. 504.

[12] To the child everything appears substantial, for the perception of which only his senses are necessary. The child asks where the shadow, where the extinguished light goes to. He will not allow the electrical machine to be turned any great length of time for fear of exhausting the supply of sparks, etc. - A boy of less than a year old wanted, when his father whistled a tune, to catch the notes from his lips. Even with older children we find the attempt to snatch at coloured after-images, etc. Only upon noting conditions of a fact that are outside ourselves does the impression of substantiality disappear. The history of the theory of heat is very instructive in this connexion.

[13] Physical experiences other than mechanical may approach to the value of mechanical experiences as they become more familiar. In my opinion Stricker has advanced a correct and important view in bringing causality into connexion with the will (*Studien über die Assoziation der Vorstellungen.*) Vienna, 1883). When I was a young lecturer in 1861, I myself advocated with great warmth and one-sidedness (in the exposition of Mill's method of difference) the view subsequently expressed by Stricker. And the idea has never quite left me (cp., for example, my *Science of Mechanics,* English trans., pp. 84, 304, 485.) However, I am at present of the opinion, as the above discussion shows, that this question is not so simple, and must be looked at from several sides, (Cp. *Wärmelehre,* 2nd ed., 1900, p. 432.)

[14] By the direction in which its hands move the watch proclaims its descent from the sun-dial and its discovery in the northern hemisphere.

[15] Thus, although my ideal of psychology is that it should be purely physiological, I should nevertheless think it a great mistake to reject so-called "introspective" psychology entirely. For self-observation is not only an important means, but in many cases is the only means of obtaining information as to fundamental facts.

[16] Cp. *Prinzipien der Wärmelehre,* 2nd ed., 1900.

[17] See *Prinzipien der Wärmelehre,* pp. 438, 459.

[18] Cp. *Wärmelehre,* p. 455; Meinong, *Hume-Studien,* Vienna, 1877; Zindler, *Beiträge zur Theorie der mathematischen Erkenntnis,* Vienna, 1889.

[19] Petzoldt, "Das Gesetz der Eindeutigkeit," *Vierteljahresschrift für wissensehaftliche Philosophie. Vol.* XIX., pp. 146 sqq.

[20] Mach, *History and Root of the Principle of the Conservation of Energy,* translated by P. E. B. Jourdain, Chicago, Open Court Publishing Co., 1911.

[21] My recently published book, *Erkenntnis und lrrtum* (1905), contains further discussions of the question. See, in particular, pp. 426-440.

XV. How My Views Have Been Received

1.

WHEN the first edition of this book was published, opinions about it were greatly divided. But in the great majority of cases, it was points of detail that found acceptance, in so far as the reception was favorable, -while the fundamental views which had led to the details were for the most part rejected. All the public criticism that I have seen has preserved a tone of moderation, even when it has been hostile, and, in its outspokenness, has been extremely instructive to me. [1]

There is no mistaking the favorable influence which the later publications of Richard Avenarius have exercised on the estimation of my book. It surely gives much food for thought, when we find a professional philosopher establishing in an elaborate systematic treatise a position which, when taken up by a scientist, there has been a disposition to explain away as the aberration of a dilettante. To-day Avenarius' pupils, and many younger inquirers who have drawn near to my position by paths of their own, are standing at my side as allies. Nevertheless, all the critics, with few exceptions, including those who reproduce my fundamental ideas quite correctly and have certainly understood them, cannot help feeling serious objections to them. There is nothing surprising in this; for I make great demands on the plasticity of my readers. It is one thing to understand an idea logically, and another to take it up in a sympathetic spirit. The ordering and simplifying function of logic can, indeed, only begin when psychical life is in an advanced stage of development and can already boast a rich store of instinctive acquisitions. Now it is scarcely possible to attain to this instinctive pre-logical nucleus of acquisitions by logical means. It is much more a question of a process of psychical transformation, which, as I found in my own case, is difficult enough even in youth. It would therefore be too much to count on immediate agreement here. On the contrary, I am satisfied to be merely allowed a hearing at all, and to be listened to without prepossessions. I will now, following the impressions I have received from my critics, once more bring out and illustrate those points of which the reception has been most strenuously opposed. In doing this, I shall treat the objections that have been urged as typical objections, with nothing captious or personal about them, and I shall therefore not mention any names.

2.

Unless we subject ourselves to a certain compulsion, we see the earth as standing still, and the sun and the fixed stars in motion. This way of looking at the matter is not merely sufficient for ordinary practical purposes, but is also the simplest and most advantageous. But the opposite view has established itself as the more convenient for certain intellectual purposes. Although both are equally correct and equally well-adapted to their special purpose, the second view only succeeded in gaining acceptance after a severe combat with a power hostile to science, a power which in this case was in alliance with the instinctive conceptions of ordinary people. But to ask that the observer should imagine himself as standing upon the sun instead of upon the earth, is a mere trifle in comparison with the demand that he should consider the Ego to be nothing at all, and should resolve it into a transitory connexion of changing elements. It is true that on various sides, the way has long been prepared for this conception. [2] We see such unities as we call "I" produced by generation and vanishing in death. Unless we indulge ourselves in the fiction, so fantastic nowadays, that these unities existed before birth in a latent state, and will similarly continue to exist after death, we can only suppose that they are just temporary unities. Psychology and psycho-pathology teach us that the Ego can grow and be enriched, can be impoverished and shrink, can become alien to itself, and can split up, - in a word, can change in important respects in the course of its life. In spite of all this, the Ego is what is most important and most constant for my instinctive conceptions. It is the bond that holds all my experiences together, and the source of all my activity. In just the same way, again, a rigid body is something very constant for our crude instinctive conceptions. If it is divided, dissolved, or chemically combined with another body, the number of these constancies increases and diminishes. Then, in order to hold fast at any price to the notion that has become so dear to us, we assume latent constancies, and take refuge in atomism. Inasmuch as we are often able to restore again the body which has disappeared or changed, this procedure rests upon somewhat better grounds than in the case of the Ego.

Now in practice we can as little do without the Ego-presentation when we act, as we can do without the presentation of a body when we grasp at a thing. Physiologically we remain egoists and materialists, just as we always see the sun rise again. But theoretically this way of looking at the matter cannot be maintained. Let us change it by way of experiment. If in doing so we obtain a glimpse of the truth, it will in the long run bear practical fruits as well.

3.

Anyone who has at some time or another been influenced by Kant, - anyone who has adopted an idealistic standpoint, and has been unable to get rid of the last traces of the notion of the "thing in itself," retains a certain inclination towards solipsism, which will appear more or less clearly. Having been through it in my early youth, I know this condition of mind well, and can easily understand it. The philosophical thinker proceeds to make the single problem of the Ego, - a problem which is in principle insoluble, - the starting-point for everything else. The Ego is something given to us, we cannot transcend it and get away from it. When, therefore, speculative philosophers say "Solipsism is the only logically consistent standpoint," their utterance is quite intelligible in view of their struggle to reach a closed, all-inclusive, complete system of the universe. To be sure, we ought to add that materialism also is equally consistent for anyone who believes that matter is the only thing that is immediately given, and that cannot be further explained. This, indeed, is true of all systems. But when a man of science tells me that solipsism is the only consistent standpoint, he excites my astonishment. I will not emphasize the fact that this standpoint is better suited to a fakir who dreams his life away in contemplation, than to a serious, thoughtful and active man. But what I do believe is that the man of science who inclines this way is making a confusion between philosophical and scientific methods. The man of science is not looking for a completed vision of the universe; he knows beforehand that all his labor can only go to broaden and deepen his insight. For him there is no problem of which the solution would not still require to be carried deeper; but there is also no problem which he can regard as absolutely insoluble. If it is impossible for the time being to make any impression on a problem, he solves in the meanwhile others that are more accessible. If he then returns to the original problem, it has generally lost much of its terrifying appearance.

No doubt the Ego is not exhausted, if we say, quite provisionally, that it consists in a peculiar connexion of the elements, as long as the nature of this connexion is not investigated in detail. But the special problems that are relevant here will not be solved by speculation j their solution will be found by the psychologists, physiologists, and psychiatrists, to whom we already owe many important elucidations of such problems. The physical substratum of the Ego, the body, [3] will afford many points of reference which introspective psychology can only handle in a very imperfect manner. A man of science who should be a solipsist would be like a physicist for whom the thermometer was the fundamental problem of the universe, because on any particular day he did not happen to have a perfectly clear understanding of the influence of temperature on expansion. On the other hand, the philosopher who is a solipsist seems to me to be like the man who gave up turning round because whatever he saw was always in front of him. As to the instinctive, but untenable, splitting up of the Ego into an object experienced and an active or observing subject, - a problem which has tormented everybody long enough, - anyone who wishes to think out these questions may compare in Ch. I.

4.

Whoever cannot get rid of the conception of the Ego as a reality which underlies everything, will also not be able to avoid drawing a fundamental distinction between my sensations and your sensations. In the same way, whoever believes in the absolute constancy of a body, thinks of this body as the single vehicle of all its properties. But when this silvery-white piece of sodium is melted, and dissolves in steam which looks absolutely different from the original thing; when the sodium is divided into different parts and transferred to different chemical combinations, so that more, or even fewer, bodies are present than before; then our habitual manner of thought can only be preserved by extremely artificial devices. It then becomes more advantageous to regard the particular properties as belonging sometimes to one and sometimes to another complex, or body, and to substitute, for the bodies that are not constant, the law which is constant and which survives the change of the properties and of their connexions. Here again, it is making no small demand, to ask that this new habit of thought should be adopted. How the thinkers of antiquity would have protested, if someone had said to them, "Earth, water and air are not constant bodies at all; what are constant are the modern chemical elements of which they are composed, many of which elements cannot be seen, while others can only be isolated or fixed with great difficulty. Fire is not a body at all, but a process," and so on. We are scarcely able to estimate correctly nowadays the magnitude of the change which lies in this step. Yet in modern chemistry a further transformation in this direction is being prepared, and the same methods of abstraction lead in due course to the standpoint which is adopted here. From the standpoint which I here take up for purposes of general orientation, I no more draw an essential distinction between my sensations and the sensations of another person, than I regard red or green as belonging to an individual body. The same elements are connected at different points of attachment, namely the Ego's. But these points of attachment are not anything constant. They arise, they perish, and are incessantly being modified. But where there is no connexion at a given moment, there is also no perceptible reciprocal influence. Whether it may or may not prove possible to transfer someone else's sensations to me by means of nervous connexions, my view is not affected one way or the other. The most familiar facts provide a sufficient basis for this view.

5.

Perhaps even more than in my fundamental ideas, many readers have found a stumbling-block in what they took, erroneously indeed, to be the general character of my conception of the universe. And, to begin with, I must say that anyone who, in spite of repeated protests from myself and from other quarters, identifies my view with that of Berkeley, is undoubtedly very far removed from a proper appreciation of my position. [4] This misconception is no doubt partly due to the fact that my view was developed from an earlier idealistic phase, which has left on my language traces which are probably not even yet entirely obliterated. For, of all the approaches to my standpoint, the one by way of idealism seems to me the easiest and most natural. And connected with this is the fear of pan-psychism, which at the same time seizes my readers. Many are the victims that fall a prey to pan-psychism, in the desperate struggle between a monistic conception of the universe and instinctive dualistic prejudices. In my early youth I had to work through these tendencies myself, and Avenarius was still labouring at them in his book of 1876. As regards these two points, I feel it to be a piece of particularly good fortune that Avenarius has developed the same conception of the relation between the physical and psychical on an entirely realistic, or, if the phrase be preferred, a materialistic foundation, so that I need do no more than refer to his discussions.

6.

My world of elements, or sensations, strikes not only men of science, but also professional philosophers, as too unsubstantial. When I treat matter as a mental symbol standing for a relatively stable complex of sensational elements, this is described as a conception which does not make enough of the material world. The external world, it is felt, is not adequately expressed as a sum of sensations; in addition to the actual sensations, we ought at least to bring in Mill's possibilities of sensation. In reply to this, I must observe that for me also the world is not a mere sum of sensations. Indeed, I speak expressly of functional relations of the elements. But this conception not only makes Mill's "possibilities" superfluous, but replaces them by something much more solid, namely the mathematical concept of function. Had I ever dreamt that a short, precise expression would be so easily overlooked, and that a popular exposition on broad lines would have been more useful, some such exposition as that which H. Cornelius [5] has so admirably given "on the concept of objective existence," would have served my purpose. In any case, even here I should have avoided the expression "possibility," and should have substituted for it the concept of function.

From expressions used in other quarters, it would appear that the true explanation of my position is to be sought in an exaggerated sensationalism, and in a correspondingly inadequate understanding of the value of abstraction and conceptual thought. Now, without a fairly well-marked sensationalism a man of science cannot accomplish much. But this does not prevent him from forming clear and precise concepts. On the contrary. The concepts of modern physics will stand comparison, in point of precision and height of abstraction, with those of any other science; but they offer at the same time the advantage that they can always be traced back with ease and certainty to the sensational elements on which they are built up. For science the gulf between intuitional presentation and conceptual thought is not so great, and is not unbridgeable. I may remark in passing that I am far from thinking meanly of the concepts of physics; for nearly forty years I have been occupied with the criticism of them in various ways, and with greater thoroughness than they have received before. And since my results are gradually, after long resistance, finding acceptance with physicists, it will perhaps be allowed that this is no cheap and facile agreement. When the physicist, whose training has accustomed him to having a kilogram weight pressed into his hand with every definition, gradually expresses himself as satisfied with definitions which reduce everything to a functional relation of sensational elements, the philosopher will surely not want to be even more of a physicist than the physicist. Naturally, however, there is no room for the necessary working out of details in this sketch, which is intended to be merely a programme for the closer connexion of the exact sciences with one another; for further information the reader must be referred to my works on physics. It would, indeed, be highly presumptuous of me to assume even that all physicists are acquainted with these works, much more that they are familiar to people who are not professional physicists; yet it is partly want of familiarity with my works which has made it possible for me to be accused, for instance, of having entirely overlooked the "spontaneity" and "autonomy" of thought. Even towards bare sensations our attitude is not one of mere passivity; for sensations disengage a biological reaction, of which the natural continuation is precisely the adaptation of thought to facts. If this adaptation were immediately and perfectly successful, the process would *ipso facto* come to an end. But since different imperfectly adapted thoughts come into conflict with one another, the biological process continues. What I have called the adaptation of thoughts to one another takes place. Now I should really like to know what process of scientific development, the logical process included, is not covered by this statement? Here I may be permitted to break off for the present these controversial remarks, in which I have only been forced to repeat what I have frequently said and have long been saying.

7.

To many readers the universe, as conceived by me, seems to be a chaos, a hopelessly tangled web of elements. They feel the want of leading and unifying points of view. But this depends on a misinterpretation of the task that I have set myself. All points of view, which are of value for the special sciences and for the philosophical consideration of the world, remain capable of further application, and indeed, are^ so applied by me. The apparently destructive tendency of the work is merely directed against superfluous, and therefore misleading, additions to our concepts. Thus I believe that the contrasts between the psychical and the physical, and between subjective and objective, have been correctly reduced by me to what is essential in them, and at the same time have been purged of traditional and superstitious conceptions. And this has been done in such a way that scientifically established points of view are not altered, and at the same time room is made for new points of view. I have no desire to set up, in the place of the lamentations of a piously whining "Ignorabimus," an obstinately self-sufficient attitude of rejection of everything that is worth knowing and that can be known. For to refuse to attempt answers to questions that have been recognized as meaningless, is in no sense an act of resignation; in view of the mass of material that can really be investigated, it is the only reasonable course open to a man of science. The physicist who refrains from seeking for the secret of perpetual motion, need not nowadays regard this as an act of resignation, any more than the mathematician, who no longer troubles himself about the squaring of the circle, or the solution of equations of the fifth degree in closed algebraic form. So, too, with more general philosophical questions: the problems are either solved, or are recognized as pointless.

"In what exactly does the fallacy, or the bias, of Mach's philosophical views consist?" This question, which one of my critics asks, strikes me as very harmless. For I am convinced that my exposition is full of defects in more than one direction. This, indeed, can scarcely be avoided when a writer's views are undergoing a radical process of revolution, for even a single head cannot work out such a process completely to its conclusion. Hence, though I can feel these faults, I cannot put my finger on them. If I could, I should be a long way further advanced towards my goal. But neither have I been able to obtain a clear view of my faults from the writings of my critics. Let us, therefore, wait a little longer.

Arguments have been brought against my views, which have been fully discussed both in this book and in other writings of mine; but I do not state this fact with a desire to reproach anybody. It must be a real torture to have to read everything that is published, and, what is more, to have to pass judgment conscientiously and deliberately in a brief allotted time. I have never discovered in myself any taste for this important vocation, and consequently I have only written three reviews, all told, in a period of forty years. So I do not grudge it to the reviewers, that they should have saved themselves a certain amount of trouble, even though it has been partly at my expense. I hope they will not take it ill on my part, if I do not re-act to every sally and to every sarcasm which they fancy has hit its mark.

Hönigswald, however, has subsequently devoted a book to my standpoint (*Zur Kritik der Machschen Philosophie,* Berlin, 1903). I must admit that he has taken the trouble to read my books; nor have I the least objection to make to a criticism which decides that my position is incompatible with Kant's. Not all philosophers will draw the inference that my position must therefore be untenable. My relations to Kant have been peculiar. His critical idealism was, as I recognize with the greatest gratitude, the starting-point of all my critical thought; but it was impossible for me to retain my allegiance to it. I very soon began to gravitate again towards the views of Berkeley, which are contained, in a more or less latent form, in Kant's writings. By studying the physiology of the senses, and by reading Herbart, I then arrived at views akin to those of Hume, though at that time I was still unacquainted with Hume himself. To this very day I cannot help regarding Berkeley and Hume as far more logically consistent thinkers than Kant. It is not the business of a man of science to criticize or refute a philosopher like Kant, though it may be observed in passing that it would no longer be a particularly heroic achievement to show the inadequacy of Kant's philosophy as a guide to modern scientific research. This has long since been effected by the progress that has been made in all departments, including philosophy itself. When Honigswald enunciates a number of general points of view, and proceeds to elicit from them a closed philosophical system, he completely misapprehends the cautiously tentative methods of approximation employed by science. The constants of the man of science are not absolutely constant, nor, on the other hand, do the changes which he investigates correspond to the limitless flux of Herakleitos. I call biological aims "practical," when they are not directed to pure knowledge as an end in itself. Only consider what the position of the man of science would be, if, before he began to think, he had to refute all the philosophical systems one by one. Once more, there is no such thing as "the philosophy of Mach." [6]

8.

Whether I shall ever succeed in making my fundamental ideas plausible to the philosophers, I must leave to time to decide. I do not attach much importance to this at present, though I have a deep reverence for the gigantic intellectual labors of the great philosophers of all ages. But I have an honest and lively desire for an understanding with the natural scientists, and I consider that such an understanding is attainable. I should like the scientists to realize that my view eliminates all metaphysical questions indifferently, whether they be only regarded as insoluble at the present moment, or whether they be regarded as meaningless for all time. I should like them, further, to reflect that everything that we can know about the world is necessarily expressed in the sensations, which can be set free from the individual influence of the observer in a precisely definable manner (see Ch. XIV.). Everything that we can want to know is given by the solution of a problem in mathematical form, by the ascertainment of the functional dependence of the sensational elements on one another. This knowledge exhausts the knowledge of "reality." The bridge between physics, in the widest sense, and scientific psychology, is formed of these very elements, which are physical and psychical objects according to the kind of combination that is being investigated.

9.

Probably a good many physiologists have taken objection to a point of detail in my position, as to which I should like to say something more. I have a great value for researches such as those of S. Exner, [7] and I believe that many important problems as to psychical phenomena can be solved merely by the investigation of the nervous connexions of the central organs, [8] and by observation of the way in which stimuli are arranged in a quantitative scale. [9] Indeed Exner's book itself is evidence of this. But I feel that the main problems still remain unsolved. For, from my point of view, I cannot conceive, any more than I could nearly forty years ago, how the qualitative variety of sensations can arise from the variation of the connexions and from mere quantitative differences. Fechner's psychophysics, which have had so important an influence, did not fail to stimulate me exceedingly at the time. Inspired by Fechner's book, I delivered some very bad lectures on the subject, the value of my lectures being still further diminished by the fact that I soon came to see that Fechner's theory of formulae of measurement was erroneous. In this connexion, after explaining Helmholtz's "telegraph-wire" theory of sensation, I said: "But will the electric processes in the nerves prove to be too simple to explain adequately the difference of quality in sensations? Will it be necessary to thrust the explanation further back into regions that are still unknown? What if, after investigating the whole brain, we find everywhere nothing but electric currents? My personal opinion is this. The electrical researches that have been made on the nerves are no doubt of a very delicate nature, but in one respect they are very rough. An electric current of given intensity tells us nothing, except that a definite quantity of living force passes in the time-unit through a cross-section of the current. By what processes and by what molecular movements that living force is assisted, we do not know. It is possible that the most diverse processes underlie one and the same intensity of current." [10] Even to-day I have not succeeded in getting rid of this idea, and I cannot refrain from bringing forward evidence that confirms it in essentially the same form, as for instance by referring to the presence of an identical current in different electrolytes. [11] The progress of physiological chemistry, [12] and the experiments that have been made in the transplantation of different organs, [13] seem to me to-day to be still more decisively in favor of my view. Rollett [14] has brought into connexion with one another, and discussed in a very instructive manner, with reference both to his own work and to that of others, a number of important questions closely related to the discussions of this book.

[1] That private judgments had been equally moderate I should not have believed, even if certain small indiscretions had not given me evidence to the contrary. A more than contemptuous judgment of a German colleague was communicated to me by a curiously roundabout path - let us say more or less by way of the Antipodes - with the unmistakable intention of giving me pain. This object, to be sure, was not attained. For it would certainly be very unfair if I were to refuse to others the right, which I exercise often enough myself, of neglecting work that I consider unprofitable. To be sure, I have never felt it necessary to insult people whose opinions differ from my own.

[2] Cp. the standpoint of Hume and Lichtenberg. For thousands of years past Buddhism has been approaching this conception from the practical side. Cp. Paul Carus, *The Gospel of Buddha,* Chicago, 1894. Cp. also the wonderful story unfolded by the same writer in *Karma, A Story of Early Buddhism,* Chicago, 1894.

[3] But what is in question here is not a transcendental, unknowable Ego, which many philosophers perhaps still think it impossible to eliminate as a last remnant of the thing-in-itself, although, generally speaking, they have risen superior to that notion by now.

[4] Shall once again state the difference in a word? Berkeley regards the "elements" as conditioned by an unknown cause external to them (God); accordingly Kant, in order to appear as a sober realist, invents the "thing-in-itself"; whereas, on the view which I advocate, a dependence of the "elements" on one another is theoretically and practically all that is required. It seems to me that, in the interpretation of Kant, his very natural and psychologically intelligible fear of being considered fantastic, has not been sufficiently taken into account. It is only from this point of view that we can understand how, while holding that only those concepts had meaning and value which were applicable to a possible experience, he could posit a thing in itself, of which no experience is conceivable. Over against the particular sensation, the plain man and the man of science both set the thing as a presentational complex of all the experiences, whether remembered or still expected, which are connected with the sensation in question; and this procedure is extremely shrewd. But for anyone who has assimilated Kant's way of thinking, it becomes meaningless at the limits of experience.

[5] *Psychologie als Erfahrungswissenschaft,* Leipzig, 1897, p. 99, and particularly pp. 110 and 111.

[6] Cf. *Erkenntnis und Irrtum,* 1905, Preface.

[7] *Entwurf zu einer physiologischen Erklärung der psychischen Erscheinungen,* Vienna, 1894.

[8] *Entwurf zu einer physiologischen Erklärung der psychischen Erscheinungen,* p. 4, Vienna, 1894.

[9] *Op. cit.* p. 3.

[10] "Vorlesungen über Psychophysik," *Zeitschrift für praktische Heilkunde,* pp. 335, 336, Vienna, 1863.

[11] See the preface to the preceding English edition of this book, Chicago, 1897, pp. v, vi.

[12] Huppert, *Ueber die Erhaltung der Arteigenschaften,* Prague, 1896.

[13] Ribbert, "Ueber Transplantation von Ovarium, Hoden, und Mamma," *Archiv für Entwicklungsmechanik,* 1898, Vol. VII.

[14] "Entwicklungslehre und spezifische Energie," *Mitteilungen des Vereins der Aerzte in Steiermark, 1902*, No. 8.

www.ingramcontent.com/pod-product-compliance
Lightning Source LLC
Chambersburg PA
CBHW030010290326
41934CB00005B/280